U.S. Combat Helmets of the 20th Century

Mass Production

U.S. Combat Helmets of the 20th Century

Mass Production

Mark A. Reynosa

Schiffer Military/Aviation History
Atglen, PA

Acknowledgments

This book is the result of a great deal of assistance, by both individuals and agencies. Those involved provided assistance or information that was often extremely hard to locate. Their help was greatly appreciated and without them this accurate history of Mass Produced U.S. Combat Helmets of the 20th Century could not be told.

The following are those who deserve a special thanks and who made the effort possible:

Chris Armold, Helmet Collector, Author, and Historian; Firestone Tire & Rubber Inc.; California State University, Northridge, Interlibrary Loan Office, Oviatt Library; Central Library, Interlibrary Loan Office, City of Los Angeles; Bernard M. Corona, Chief, Close Combat-Light & Heavy Division, Human Engineering Laboratory, U.S. Army Ordnance Center and School; Michel De Trez, Helmet Collector, Author, and Historian; Robert W. Fisch, Museum Curator, West Point Museum; Luther Hanson, Museum Specialist, U.S. Army Quartermaster Museum; Larry Holtzman, Souix Manufacturing; Keith R. Jamieson, M.D., Helmet Collector; Jon A. Maguire, Collector, Author, and Historian; Lisa A. Martellaro-Reynosa, Engineering Research and Support; Linda C. Martinez, Photo Support and Photo Equipment; Jack Massopust, U.S. Army Veteran; Don Mauro, Gentex Corporation; Patrick Murphy, Specialty Plastic Products; Russell P. Nardi, Manager, Advertising and Marketing Materials, and Michael J. Cameron, Manager, Marketing Communications, McCord Gaskets, AE Clevite Inc.; Clayton R. Newell, Lieutenant Colonel, U.S. Army, Chief, Historical Services Division, Center of Military History; Edward J. Reese and DeAnne Blanton, Military Reference Branch, National Archives; Paula Reed, Librarian, Mine Safety Appliances Company; Crystal A.G. Reynosa, Photo Support; Joseph G. Reynosa, U.S. Army Veteran; Michael R. Reynosa and Julie C. Reynosa, Mil-Spec Research, Engineering Research and Photo Equipment; Raymond G. Reynosa, World War II U.S. Army Veteran; Robert G. Reynosa and Olivia B. Reynosa, Research and Support; Norman C. Ristaino, Dennis Gordon, Harvey W. Keene, and A Reardon, U.S. Army Natick Reserch, Development and Engineering Center; Charles A. Ruch, Historian, Westinghouse Electric Corporation; Jerry Schilling, Unicor Corporation Divsion of Federal Prison Industries; John J. Slonaker, Chief, Historical Reference Branch, U.S. Army Military History Institute; K.L. Smith-Christmas, Material History Curator, United States Marine Corps Air-Ground Museum, Museum Branch Activities; Larry Sutherland, U.S. Air Force Veteran and Collector; Carlos Vidal, U.S. Army Veteran; Anthony C. Wilsbacher, U.S. Army Veteran, and Technical Reviewer; and a special thanks to Bob Biondi and Peter Schiffer for publishing this book!

To God, Family and Country

Book Design by Ian Robertson.

Library of Congress Catalog Number: 97-67427

Printed in China.
ISBN: 0-7643-0357-0

Published by Schiffer Publishing Ltd.
4880 Lower Valley Road
Atglen, PA 19310
Phone: (610) 593-1777
FAX: (610) 593-2002
E-mail: Schifferbk@aol.com
Please write for a free catalog.
This book may be purchased from the publisher.
Please include $3.95 postage.
Try your bookstore first.

CONTENTS

INTRODUCTION

This book culminates the series of booklet monographs that I have written over the past 10 years concerning U.S. Combat Helmets, most notably the M-1 helmet. My research into U.S. Combat Helmets began during my years of study in the school of Engineering and Computer Science, at California State University, Northridge. In the spring of 1988, at California State University Northridge, I began to research the history of the U.S. M-1 Steel Combat Helmet. Within the next two years, I had gathered an extensive amount of information on the M-1 Helmet. What began as a simple quest for information on the topic had lead to one of the largest formal efforts to accurately compile the history of the M-1 Helmet. In addition to acquiring information on just the M-1 helmet, I was also compiling a substantial amount of information on other U.S. Combat Helmets. My early research on the topic during 1988 and 1989, led me to compile a small set of notes.

These early notes where compiled into a small booklet. This small booklet was the first effort that I was aware of that tried to document the history of the M-1 Helmet. Subsequently this booklet was later distributed to others interested in the history of the M-1 Helmet. The small booklet was entitled "M1 Steel Combat Helmet; A Brief History and Visual Reference Report, 1941-1989". This was sometimes referred to as the "Green Booklet". Four revisions of this booklet were finally distributed between May 1988 and May 1990. While the first version of this booklet contained information that could not be substantiated, the forth revision had replaced all of the unsubstantiated information with substantiated researched facts.

By the winter of 1990, I had compiled enough information to write a small manuscript dealing with the history of the M-1 Helmet during World War II. Thus in January 1991, I wrote my small concise manuscript on the M-1 Helmet during World War II. In the spring of 1992 I was able to compile a small manuscript on the M-1917 and M-1917A1 Helmets. Neither of these manuscripts took the form of a booklet. By 1993 I had received numerous inquires from historians and collectors wondering whatever became of my research. So, in January 1994, I decided to manufacture a small number of copies of the M-1 Helmet manuscript as a booklet. This booklet was entitled "The M1 Helmet, 1941-1945; A Short History". This was sometimes referred to as the "White Booklet".

In the spring of 1995, Schiffer Publishing Ltd., was gracious enough to provide me with the opportunity to publish a helmet book. Thus, in 1995, the research effort to find even more information on various U.S. Helmets was back on again. The book entitled "The M-1 Helmet, A History of the U.S. M-1 Helmet in World War II" was published in the summer of 1996. Now, almost ten years after I started, I have written this Schiffer book, which along with "The M-1 Helmet, A History of the U.S. M-1 Helmet in World War II" book, culminates all of the information I have gathered on the history of U.S. Mass Production Combat Helmets.

1

THE M-1917 HELMET

The United States entered into World War I in April 1917, at this time the United States Army did not have a hel met for its troops. The adoption of a helmet by the French, British and German armies convinced the United States Army that a helmet was needed as a standard piece of equipment. In June 1917, the United States Army selected the standard British helmet design for its use. This was the British Mk. I steel helmet. There were three main reasons for the selection of the British Mk. I helmet design: "the immediate availability of 400,000 ready-made helmets from England, the simplicity of manufacture from hard metal, and the superior ballistic properties." When the British Mk. I was selected by the United States Army, its United States production version was designated and standardized as the Helmet, M-1917. Until United States production of the M-1917 could begin, the United States Army purchased the 400,000 available British Mk. I helmets in England and issued them to the American Expeditionary Forces already in Europe.

Production was begun on the M-1917 helmets in the fall of 1917. By the end of November 1917, large quantities of M-1917 helmets became available for the United States Army.

Exterior right top view of the M-1917 helmet, ca. 1917-1918. (Courtesy of U.S. Army)

The M-1917 helmet was very similar to the British Mk. I helmet. The helmet was basically an inverted bowl stamped out of a single piece of manganese alloy, which was made up of 13 percent manganese and was .036" thick. This differed from the British helmet, as the Mk. I helmet was made up of 12 percent manganese. Thus ballistically, the M-1917 helmet increased protection for the wearer by 10 percent over the British Mk. I helmet, and could withstand a .45 caliber pistol bullet traveling at 600 feet per second fired at a distance of 10 feet. A rim was spot welded to the edge of the steel bowl, with the ends butted, as opposed to lapped, which was done on the British Mk. I helmet. Riveted to the steel bowl were two flexible guiding loops for the chin strap. Here again, the U.S. M-1917 helmet differed from the British Mk. I helmet. On the U.S. helmet the loops were secured by solid machined rivets,

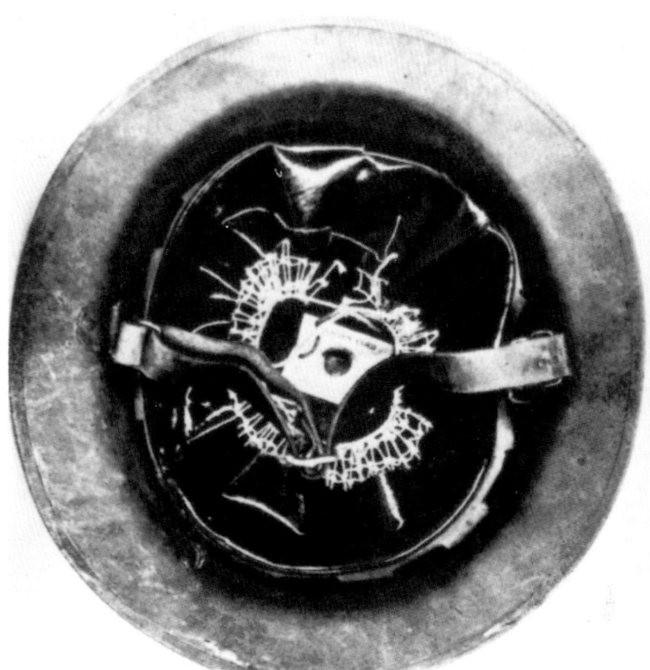

Interior view of the M-1917 helmet, ca. 1917-1918. (Courtesy of U.S. Army)

Exterior front right view of the M-1917 helmet, ca. 1917-1918.

Interior view of the M-1917 helmet, ca. 1917-1918.

M-1917 chin strap loop, ca. 1917-1918. Note the machined rivet used to secure the loop to the helmet. Also note the split rivet used to secure the leather chin strap.

whereas the British Mk. I helmet used split rivets. An adjustable leather chin strap was riveted to the steel bowl and consisted of two halves, each joined together by metal loops which were secured to the ends of the leather halves by steel split rivets. Also riveted to the steel bowl was the helmet lining. The lining was also similar to that of the British Mk. I helmet and consisted of a number of items described below:

"The lining was woven of cotton twine in meshes three-eighths of an inch square. This web, fitting tightly upon the wearer's head, evenly distributed the weight of the two-pound helmet, and in the same way distributed the force of any blow upon the helmet. The netting, together with the small pieces of rubber around the edge of the lining, kept the helmet away from the head, so that even a relatively large dent could not reach the wearer's skull."

Close-up of the M-1917 crown instruction tag, ca. 1917-1918. The instructions read "TIGHTEN CORD AND ADJUST NET TO FIT THE HEAD".

Close-up of the M-1917 inspection stamp, ca. 1917-1918. The stamp reads "ORD. DEPT. INSPTC. 117".

M-1917 chin strap buckle, ca. 1917-1918. The buckle was made of steel and left unpainted.

The linings of the U.S. M-1917 helmet were produced by 10 shoe manufacturing companies. The lining, as mentioned above, consisted of cotton twine mesh surrounded by a circular piece of leather that held tubular pieces of rubber, and the mesh was covered by a piece of black oil cloth. Sandwiched between the lining and the steel bowl was also piece of felt.

The steel for the M-1917 helmet was rolled by the American Sheet and Tin Plate Company. The steel was then pressed and stamped into its bowl shape by seven companies, which were: Edward G. Budd Manufacturing Company, Philadelphia, Pennsylvania; Sparks, Withington Company, Jackson, Michigan; Crosby Company, Buffalo, New York; Bossett Corporation, Utica, New York; Columbian Enameling & Stamping Company, Terre Haute, Indiana; Worchester Pressed Steel Company, Worchester, Massachusetts; and Benjamin Electric Company, Des Plaines, Illinois. The steel was stamped with an austenite heat number and shipment number, which were used to identify the quality of steel and shipment lots.

The metal helmets and woven linings were delivered to the plant of the Ford Motor Company, Philadelphia, Pennsylvania, were they were painted and assembled. To make the

Close-up of the M-1917 stamp marking, ca. 1917-1918. The stamp was placed under the brim of the helmet and represented a heat and shipment number.

Close-up of the M-1917 edging butt, ca. 1917-1918.

Close-up of the M-1917 exterior finish, ca. 1917-1918. This view shows the sawdust aggregate applied to the olive drab finish.

outside surface of the helmet anti-glare, the helmets were first painted, then fine sawdust was blown on the wet paint, and finally the helmet was painted again. To increase protective properties the helmets were painted in an olive drab shade.

M-1917 HELMET PROCUREMENT AND TOTALS

During the fall of 1917 production was begun on the M-1917 helmets. By the end of November 1917, the first deliveries of large quantities of M-1917 helmets were being made to the United States Army. On 17 February 1918, approximately 700,000 M-1917 helmets had been produced. As United States involvement in World War I increased, the U.S. Army placed additional orders for the M-1917 helmet. By July 1918 orders for the M-1917 helmet reached 3,000,000, in August 6,000,000, and in September 7,000,000. In November 1918, when hostilities ended and American production was ordered to cease, U.S. Manufacturers had produced a total of 2,707,237 M-1917 helmets.

Production figures for the pressed and stamped steel helmets during World War I, were as follows:

Edward G. Budd Manufacturing Co.	1,150,775
Sparks, Withington Co.	473,469
Crosby Co.	469,968
Bossett Corporation	116,735
Columbian Enameling & Stamping Co.	268,850
Worchester Pressed Steel Co	193,840
Benjamin Electric Co.	33,600

2

THE M-1917A1 HELMET

At the outbreak of World War II, the U.S. Army placed an order for the only production set of M-1917A1 helmets. On 27 November 1940, the McCord Radiator and Manufacturing Company, Detroit, Michigan was contracted by the United States Army's Rock Island Arsenal to produce 1,843,431 M-1917A1 helmets. McCord quickly went to work on the project by developing the necessary manufacturing techniques. Deliveries of the M-1917A1 started in February 1941, with monthly deliveries averaging 260,000 helmets. By May 1941, when the Ordnance Department instructed that production of the M-1917A1 helmet was to be discontinued, McCord had produced 904,020 helmets.

The M-1917A1 appeared much the same as the M-1917 helmet, except for the replacement of the lining and chin strap. The new lining, officially called "head pad assembly" and developed during 1934, consisted of a center leather head hairpad, and a circular leather head lining consisting of four leather flaps laced together. The new chin strap, developed in late 1918, consisted of two pieces of web cotton. The right web strap, which was the longer of the two, had a brass buckle at its end, and the left side had a brass hook to receive it.

In addition to the new head pad assembly and new khaki, olive drab shade no. 3, cotton chin strap, the production M-1917A1 contained further changes. These changes include the use of cork aggregate in the painting of the exterior of the helmet, and a shorter character designation for the austenite heat number and shipment number.

Exterior front left view of the production M-1917A1 helmet, ca. 1940. This helmet was produced by the McCord Radiator and Manufacturing Company.

Interior view of the production M-1917A1helmet, ca. 1940.

Close-up of the production M-1917A1 stamp marking, ca. 1940. The stamp was placed under the brim of the helmet and represented a heat and shipment number. This view also show the edging butt.

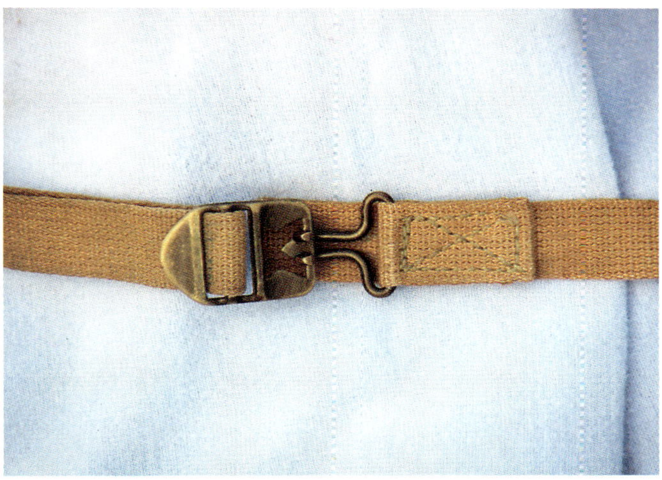

Production M-1917A1 chin strap buckle and release hook, ca. 1940.

Production M-1917A1 chin strap loop, ca. 1940. Note the machined rivet used to secure the loop to the helmet. Additionally, note the webbed chin strap attached to the suspension as opposed to the loop itself.

Close-up of the production M-1917A1 exterior finish, ca. 1940. This view shows the cork aggregate applied to the olive drab finish.

3

THE M-1 HELMET

In late 1940, the United States set out on a policy to have the best equipped army in the world. The job of developing this new helmet was given to the U.S. Army Infantry Board located at Fort Benning, Georgia. The Infantry Board had established a set of characteristics for the new helmet. In early 1941, development of the new helmet was begun.

STEEL BODY DEVELOPMENT

In January 1941, shortly after the Infantry Board had established a set of characteristics, the U.S. Army Ordnance Department officially began to develop a new "pot-shaped" helmet. The Ordnance Department acquired an assortment of various foreign combat helmets for ballistic and metallurgical testing, as well as requesting assistance from personnel at the Metropolitan Museum of Art in New York. The museum was influential in the design and development of the new helmet

After extensive research it was determined that the M-1917 helmet was the most suitable helmet to protect the top of the head. Thus, it was decided to base the new helmet on the M-1917 crown with brim removed and sides extended.

On 7 February 1941 the McCord Radiator and Manufacturing Company of Detroit, Michigan, was awarded a contract by the Ordnance Department to produce sample dies and sets of sample helmet bodies and liners. McCord was requested to make this set of sample helmet bodies using Hadfield Manganese Steel. The same steel used in the production of the M-

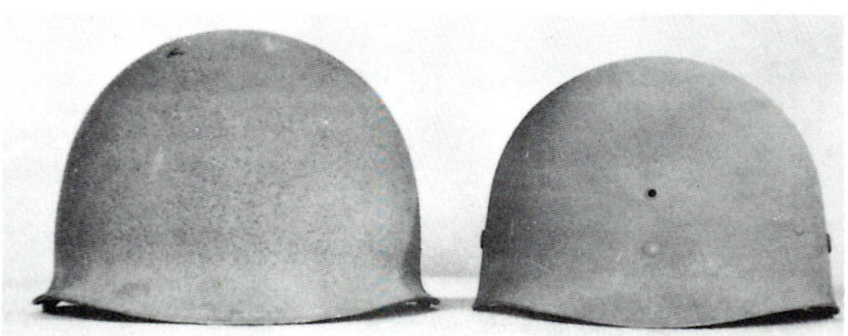

Exterior front view of the M-1 helmet and liner, ca 1943. (Courtesy of Naval Medical Research and Development Command)

Exterior side view of the M-1 helmet and liner, ca 1943. (Courtesy of Naval Medical Research and Development Command)

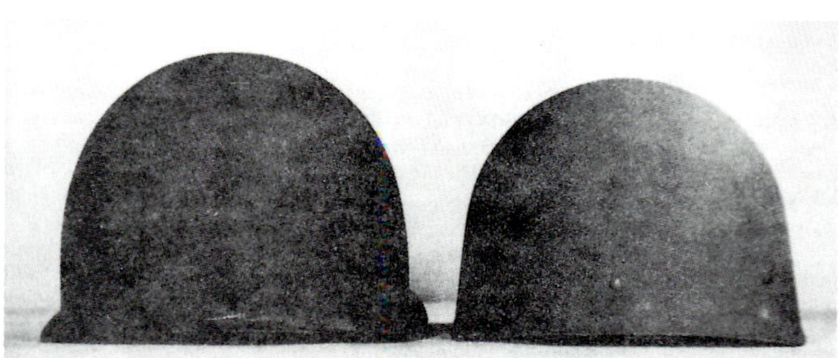

Exterior rear view of the M-1 helmet and liner, ca 1943. (Courtesy of Naval Medical Research and Development Command)

Interior view of the M-1 helmet and liner, ca 1943. (Courtesy of Naval Medical Research and Development Command)

A World War II U.S. Army soldier wearing the M-1 helmet. Note the fixed chin strap loop helmet body, ca. 1941-1943. (Courtesy of U.S. Army via Larry Sutherland)

Exterior front right view of the M-1 helmet body, ca. July 1941-October 1943.

Interior view of the M-1 helmet body, ca. July 1941-October 1943.

1917A1. This new helmet body required a 7" draw. In March 1941, McCord began the manufacture of the 200 sample helmet bodies.

LINER DEVELOPMENT

The development of the liner for the steel body was begun in early 1941. The liner for the new helmet had its origins in the Riddell plastic football helmet and suspension. The Riddell football helmet was being used in the training of the parachute troops, at Fort Benning. Also located at Fort Benning was the U.S. Army Infantry Board which became aware of the football helmet and its possiblity of serving as a liner for the new helmet. As a result, a modified form of the Riddell football helmet was used as the basis for the liner of the new helmet. Althougth the Riddell based plastic liner proved to be too heavy for practical use, the suspension was found to be superior in performance and was accepted by the U.S. Army.

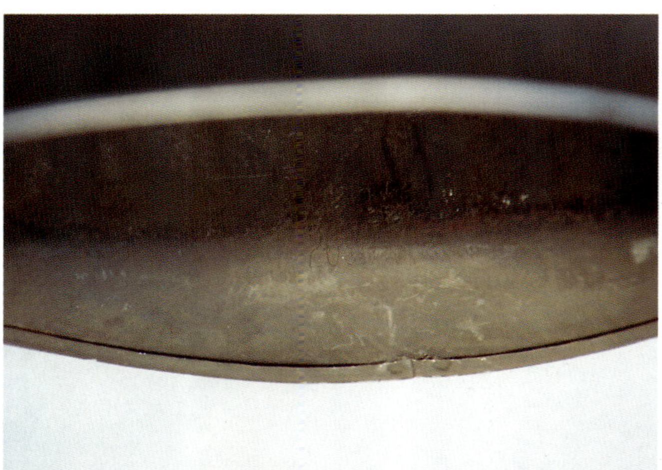

Close-up of the M-1 helmet body stamp marking. The stamp was placed under the visor and represented the heat number and the shipment number.

M-1 helmet body fixed chin strap loop, ca. July 1941-October 1943.

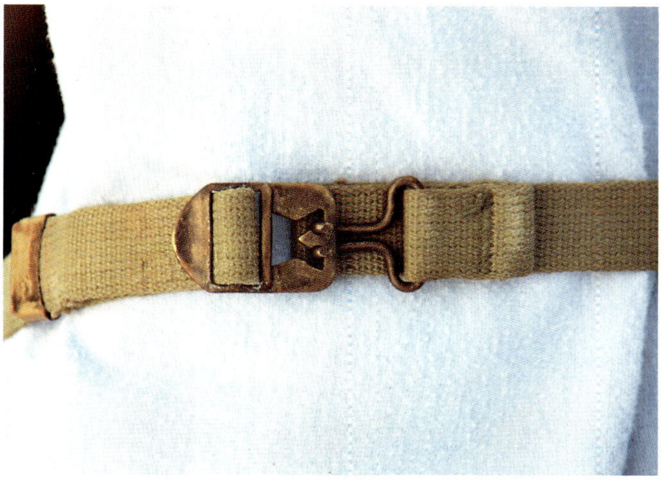

M-1 helmet body chin strap buckle and release hook, ca. 1941-1943. This variation was made of brass.

Exterior view of the M-1 helmet body front edging butt, ca. July 1941-October 1943.

During February 1941, a plastic alternative for the helmet liner was found in the Hawley tropical fiber helmet, to which the Riddell suspension system was attached. The Hawley Products Company, along with McCord and others completed the manufacture of a set of 200 sample liners. These sample liners were approved by the U.S. Army in late April 1941.

MODEL TS-3

This new combination of helmet body and liner was then designated as the Model TS-3. The TS-3 was designed to be manufactured in one size that would fit all head sizes. The TS-3 helmet then underwent extensive testing which showed that the new helmet not only provided greater ballistic protection but that the extended areas at the sides and back, being very stable on the wearer's head, did not interfere with the firing of weapons, did not obscure the wearer's field of view, and met ballistic requirements . The TS-3 had met most of the characteristics established by the Infantry Board. Thus in April 1941, the TS-3 was standardized, and on 9 June 1941 the standardization of the experimental Helmet, TS-3 as the Helmet, Steel, M-1, was approved.

Exterior view of the M-1 helmet body exterior finish, ca. 1941-1945. Note the cork aggregate applied to the olive drab exterior paint.

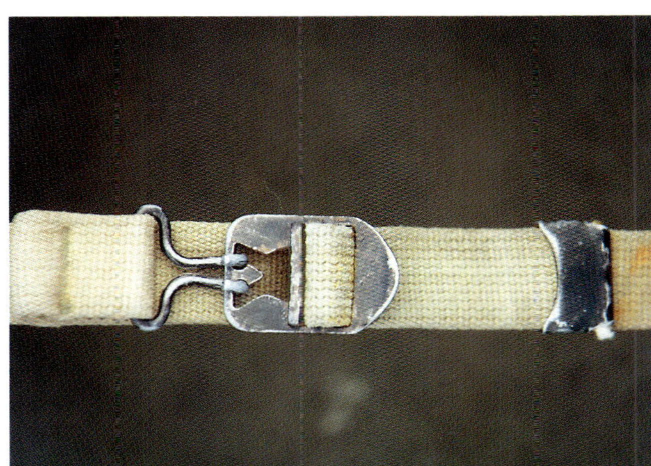

Fastened M-1 helmet body chin strap buckle and release hook, ca. 1942-1943. This variation was made of steel painted black.

M-1 helmet body with flexible chin strap loops, ca 1943-1945. (Courtesy of U.S. Army Ordnance Department via National Archives, RG 156 Ordnance.)

Exterior front left view of the M-1 helmet body, ca. October 1943-November 1944.

M-1 HELMET BODY PRODUCTION, 1941 - 1945

Production of the steel helmet body was begun in the summer of 1941 by the McCord Radiator and Manufacturing Company of Detroit, Michigan. McCord officially received its first M-1 helmet production contract on 26 June 1941. Initially production was slow, with approximately 300,000 helmet bodies being produced in 1941. By 1942, production had increased to about 5,000,000.

Interior view of the M-1 helmet body, ca. October 1943-November 1944.

M-1 helmet body flexible chin strap loop, ca. October 1943-August 1945. Note sewn on chin strap.

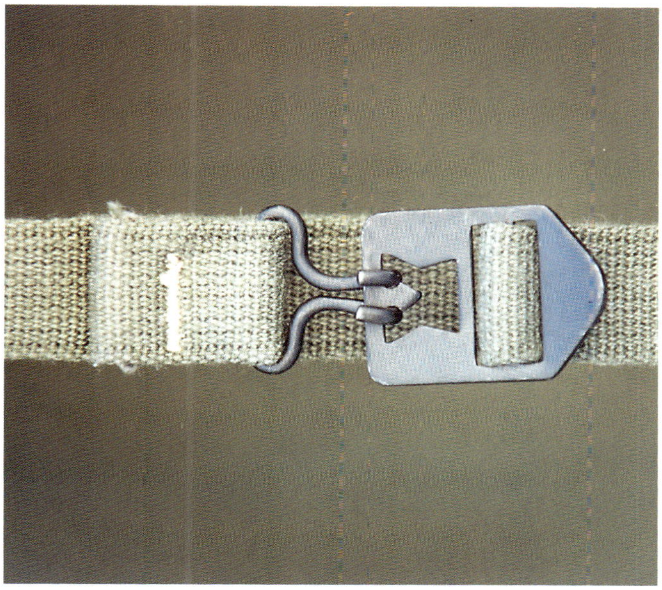

Fastened M-1 helmet body chin strap buckle and release hook, ca. 1943-1945. This variation was stamped of steel and painted black. Other variations were stamped of brass and coated with a mildew inhibitor. Note olive drab shade no 7 webbing used from November 1944 onwards.

Exterior front right view of the M-1 helmet body, ca. November 1944-August 1945.

The helmet body was manufactured from a single piece of Hadfield Manganese steel. The single pieces of steel were circular blanks, which were stamped with a heat number and shipment number that were used to identify the quality of steel and shipment lots throughout the manufacturing process.

The initial helmet body was manufactured from this single circular blank, which was formed by a single draw and then trimmed. To this shell a thin piece of stainless steel was applied by spot welding to form a rim, with the rim butting in the front of the helmet body. Also welded to the helmet were two stainless steel fixed chin strap loops. The choice to use stainless steel was made in order to allow the entire helmet to be non-magnetic. This was necessary, so that the helmet would not cause an error in compass readings during field operations.

Interior view of the M-1 helmet body, ca. November 1944-August 1945.

Interior view of M-1 helmet body rear edging butt, ca. November 1944-August 1945.

U.S. Army soldiers wearing their M-1 helmets and liners, Normandy, France, summer 1944. The soldier on the right is Raymond G. Reynosa.

Rear view close-up of the M-1 helmet body chin strap flat end clip, ca. 1941-1945. Variations were made of either steel painted black or brass coated with a mildew inhibitor. Note olive drab shade no. 7 webbing used from November 1944 onwards.

Exterior front left view of the M-1 helmet body, ca. 1951-1960.

Once the helmet body's metal parts were assembled, the helmet was placed in an acid wash to clean it prior to the application of paint. After cleaning it was painted a shade of olive drab both inside and out. During the painting process the paint was mixed with cork and applied to the outer surface so that the helmet body would have a rough, non-reflective appearance. Finally, a two piece khaki, olive drab shade no. 3, cotton web chin strap was sewn to the fixed loops. The metal components of the chin strap were each made of brass and coated with a mildew inhibitor.

The completed M-1 helmet body, manufactured in only one size, weighed approximately 2.25 pounds and measured about 0.037" thick, 9.4" wide, 11.0" long and 6.9" deep.

The initial version of the M-1 helmet body suffered from a number of minor defects. One of the problems was related to the manufacture of the initial helmets. The need to manufacture the M-1 body in a single 7" draw caused a serious problem. Early models suffered various cracks in the helmet, either caused by the initial manufacture or field use. Additionally, the helmet body had a tendency to rust due to a chemical reaction of the paint and cork. Solutions were worked out for both of these problems.

An additional problem involved the design of the rigid non-flexible chin strap loops which tended to break off during repeated field use. To solve this problem, a new flexible, hinged chin strap loop, was introduced. The new loop was also con-

Interior view of the M-1 helmet body, ca. 1951-1960.

Exterior view of M-1 helmet body rear edging butt, ca.1951-1960. Note the sand aggregate applied to the exterior paint.

M-1 helmet body flexible chin strap loop and chin strap fastener (clip), ca. 1951-1960.

Fastened M-1 helmet body chin strap buckle and release hook, ca. 1951-1960. The buckle was stamped of steel and painted light olive drab.

structed of stainless steel and was initially attached by two spot welds, and later three, to secure it to the helmet body. The previous method of sewing on the web chin strap was still acceptable. Around October 1943, production of the M-1 helmet bodies with the new hinged chin strap loop was begun.

Initially, the M-1 helmet body was manufactured with a stainless steel rim, however this later proved to be to the disliking of the Army. The Army had noticed that paint did not adhere well to this surface and could potentially cause unnec-

Rear view close-up of the M-1 helmet body chin strap flat end clip, ca. 1951-1960. The flat end clip was made of steel painted light olive drab.

Fastened M-1 helmet body chin strap buckle and release hook, ca. 1961-1965. The buckle was stamped of brass coated with a mildew inhibitor.

U.S. Army soldiers drilling and wearing M-1 helmets at Fort Ord, California, 1961-1962. The soldiers wore a mix of helmet bodies. The soldier circled is Robert G. Reynosa. (Courtesy of U.S. Army)

Exterior front left view of the M-1 helmet body, ca. 1961-1965.

essary reflection in the field. This resulted in a production change in October 1944, which saw the rim manufactured from Hadfield Manganese. Additionally, it is probable that another production change or substitution was also enacted, which required the butt of the rim to be placed at the rear of the helmet body, as opposed to the front of the helmet body.

During production, substitutions and other minor changes were allowed by the U.S. Army for the chin strap. The substitutions which occurred were the result of shortages of brass during the years of 1942 and 1943. In place of brass, the chin strap flat end clip, buckle and release hook were sometimes made of metal and given a coat of black paint. The only other change that occurred during the production of the helmet body's chin strap was the addition of a second webbing color, which was thought to have been olive drab shade no. 7, a dark green color. This change occurred in the fall of 1944.

Interior view of the M-1 helmet body, ca. 1961-1965.

M-1 helmet body flexible chin strap loop and chin strap fastener (clip), ca. 1961-1965.

M-1 helmet body chin strap flat end clip, ca. 1961-1965.

Fastened M-1 helmet body chin strap buckle and release hook, ca. 1961-1965. This variation was stamped of brass and coated with a mildew inhibitor.

M-1 HELMET BODY PROCUREMENT AND TOTALS, 1941 - 1945

Initially the U.S. Army had selected a single contractor to produce the steel helmet body, the McCord Radiator and Manufacturing Company of Detroit, Michigan. By January 1943, a second company was producing M-1 helmet bodies, the Schlueter Manufacturing Company of St. Louis, Missouri.

Production figures for the helmet bodies during World War II, were as follows:

Year	Quantity
1941	323,510
1942	5,001,384
1943	7,648,880
1944	5,703,520
1945	3,685,721

The total production of World War II M-1 helmet bodies reached 22,000,000 by VJ day, and ceased shortly thereafter. McCord manufactured approximately 20,000,000 helmet bodies, while Schlueter produced the balance of the helmet bodies.

M-1 HELMET BODY PRODUCTION, 1951 - 1960

In 1951, production of the M-1 helmet resumed. This new production version of the M-1 helmet incorporated all of the changes and recommendations brought forth from experience gained with the World War II M-1 helmet. While this new production version of the M-1 helmet resembled its forebears, it differed in production items and features.

Between 1951 and 1960 the production helmet appeared as follows: The helmet body's exterior paint was mixed with sand aggregate as opposed to cork, which was used during World War II production. The helmet body's exterior paint was

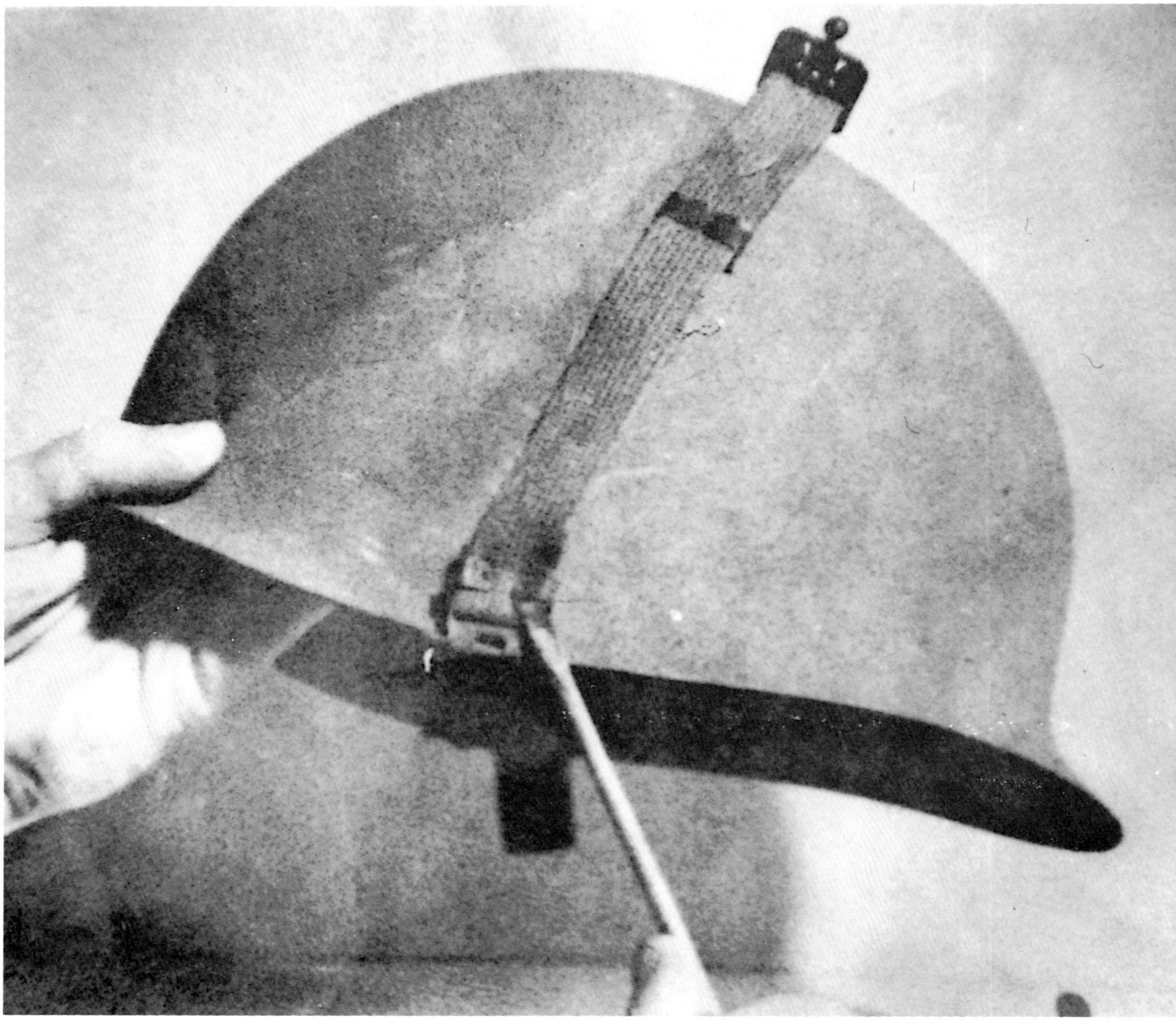

The M-1 helmet body, ca. 1966-1972. (Courtesy of U.S. Army)

Exterior front right view of the M-1 helmet body, ca. 1966-1972. (Courtesy of Larry Sutherland)

Interior view of the M-1 helmet body, ca. 1966-1972. (Courtesy of Larry Sutherland)

also slightly lighter in color than the World War II production helmet body. The 1951-1960 production helmet body color was olive drab color no. 319. The helmet body also incorporated the T1 chin strap release and T1 chin strap fastener as production items. The items were previously non-production items, accessories and components developed during the later part of World War II, and did not see much field use. The production chin strap fastener was made of steel painted a light olive drab, as were chin strap buckle and flat end clip. This appears to be a substitution, as U.S. specifications indicated that the production chin strap buckle and flat end clip were to be made of

Fastened M-1 helmet body chin strap buckle and release hook, ca. 1966-1972. This variation was stamped of steel and painted black. (Courtesy of Larry Sutherland)

M-1 helmet body flexible chin strap loop and chin strap fastener (clip), ca. 1966-1972. (Courtesy of Larry Sutherland)

M-1 helmet body chin strap flat end clip, ca. 1966-1972. Note the squared ends. (Courtesy of Larry Sutherland)

Close-up of the M-1 helmet body stamp marking. The stamp was placed under the visor and represented the heat number and the shipment number.

brass coated with a mildew inhibitor. U.S. specifications indicated that the production chin strap fastener (clip) was to be produced with a black finish.

M-1 HELMET BODY PRODUCTION, 1961 - 1965

Between 1961 and 1965 the production helmet body incorporated only one minor change. The production chin strap buckle and the flat end clip were made of brass coated with a mildew inhibitor, while the chin strap clip was made of steel painted a black. The production chin strap clip, and flat end clip had

Exterior front left view of the M-1 helmet body, ca. 1973-1988.

Front view ot the M-1 helmet body, ca. 1973-1988. (Courtesy of U.S. Army)

Side view of the M-1 helmet body, ca. 1973-1988. (Courtesy of U.S. Army)

Interior view of the M-1 helmet body, ca. 1973-1988.

M-1 helmet body flexible chin strap loop and chin strap fastener, ca. 1973-1988.

their ends enlarged to 0.375" so as to provided improved fastening to the cotton webbed chin strap.

M-1 HELMET BODY PRODUCTION, 1966 - 1972

Between 1966 and 1972 the production helmet body again incorporated only minor changes. Brass was cancelled from production for all items related to the helmet body. The production flat end clip had its ends changed from round to square. In addition the production chin strap buckle and flat end clip were produced from corrosion resistant steel with a black finish. In 1968 the production paint color of the helmet body was changed from olive drab color no. 319 to Munsell 10Y. Sand continued to be retained as the paint aggregate.

M-1 HELMET BODY PRODUCTION, 1973 - 1988

In 1973 the final production version of the M-1 helmet body saw a change in the helmet body's chin strap. The original design chin strap was replaced in production by a webbed cotton-nylon chin strap with a webbed chin cup. This new webbed cotton-nylon chin strap with a webbed chin cup was also is-

Exterior and interior views of the M-1 fiber helmet liner, ca. 1942. (Courtesy of U.S. Army Ordnance Department via National Archives, RG 156 Ordnance.)

Exterior right view of the M-1 fiber helmet liner, ca late 1941-September 1942.

Interior view of the M-1 fiber helmet liner, ca late 1941-September 1942.

sued in kit form for replacement on existing issued M-1 helmet bodies. Finally in 1988, U.S. specifications called for the paint color to be changed from Munsell 10Y to olive drab 34087.

M-1 FIBER LINER PRODUCTION, 1941 - 1942

The first fiber liners were procured along with the first order of the M-1 helmet bodies. The prime contractor was the McCord Radiator and Manufacturing Company. McCord subcontracted the fiber liner body to its co-developer, the Hawley Products Company. While Hawley manufactured the fiber liner body, McCord supplied and installed the suspension to finish the liner at the McCord plant.

In October 1941, specifications for the fiber liner were written. The specifications stated that the liner was to be made of "two shells, each a one piece rigid fiber form, impregnated with varnish or other water insoluble and water repellent materials, securely cemented together with a suitable thermoplastic or thermosetting material which shall be insoluble in wa-

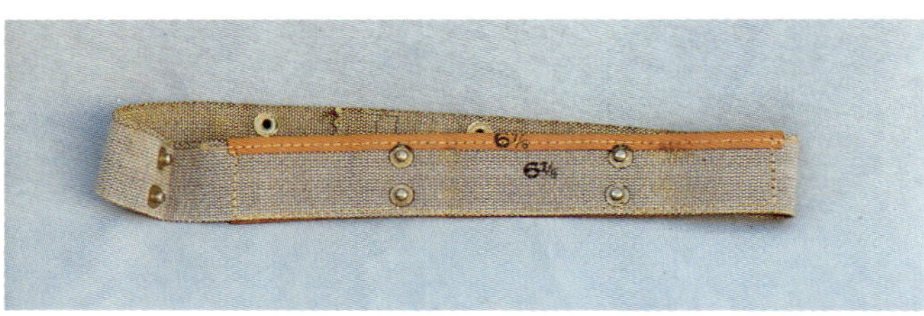

M-1 helmet liner sized head band, ca. 1941-1942. This variation the head band was made of rayon webbing.

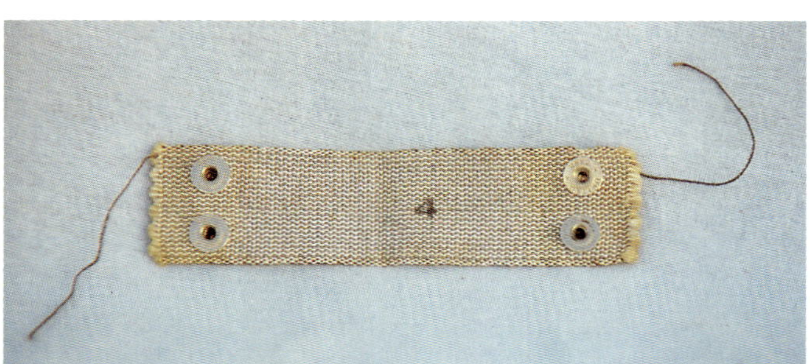

M-1 helmet liner neck band, ca 1941-1942. This variation of the neck band was made of rayon webbing.

Exterior left view of the M-1 fiber helmet liner, ca September 1942-November 1942.

Interior view of the M-1 fiber helmet liner, ca September 1942-November 1942.

Exterior front left view of an M-1 high-pressure plastic helmet liner, ca. May 1942-September 1942. This particular helmet was manufactured by the Westinghouse Electric Company.

ter." Once this body was formed, the fiber liner had a piece of olive drab gaberdine or twill smoothly cemented over it. The liner was designed and manufactured so that it would slip-fit into the M-1 steel helmet body. Three fittings were attached to this fiber liner body and included a suspension, neck strap, and an adjustable chin strap. The suspension and neck strap were both made of rayon webbing, and secured to the liner by means of perpendicular-sided metal washers. Added to this liner were two inserts, the head band and the neck band. The head band was produced in thirteen sizes, constructed of rayon webbing and had a piece of leather sewn to the front half of it. The neck band was also constructed of rayon webbing and was produced in three sizes. Both inserts were attached to the suspension and neck strap respectively, by means of snap fasteners. The leather

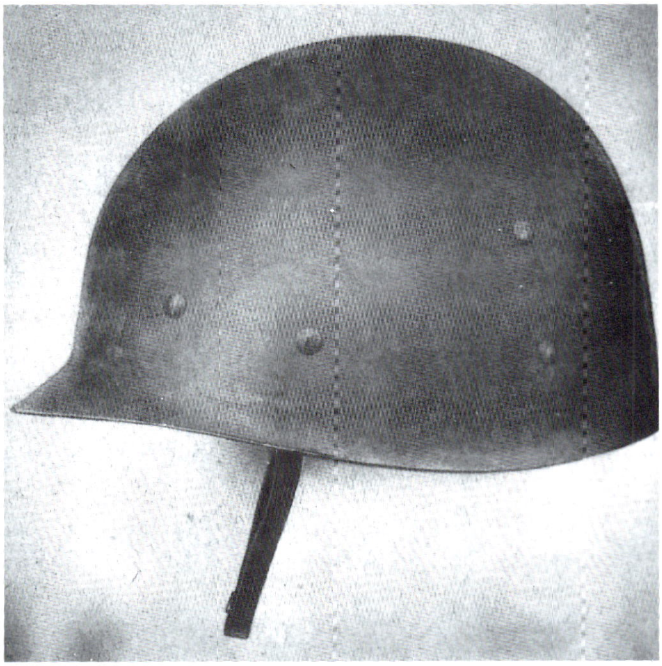

Exterior view of M-1 plastic helmet liner, ca. 1942. (Courtesy of U.S. Army Ordnance Department via National Archives, RG 156 Ordnance.)

Interior view of M-1 plastic helmet liner, ca. 1942. (Courtesy of U.S. Army Ordnance Department via National Archives, RG 156 Ordnance.)

Interior view of an M-1 high-pressure plastic helmet liner, ca. May 1942-September 1942. This particular helmet was manufactured by Westinghouse Electric Company.

Exterior left front view of an M-1 high-pressure plastic helmet liner, ca. early 1943-June 1944. This particular helmet was manufactured by the Mine Safety Appliances Company.

chin strap was adjusted by a square two slit type buckle made of metal, and was secured to the liner by permanent chin strap holders, which were riveted to the fiber liner body.

The fiber liner with the rayon webbing displayed very few markings. The only markings that did appear, were black numbers, indicating the size, which were stamped onto the head band and neck band inserts. Occasionally, the liner was marked on the inside with the following: "LINER, FIBRE M-1".

By June 1942, a new type of webbing had been developed to replace the rayon webbing in the suspension system. This and other changes which are discussed in further detail later in this chapter, resulted in approximately the last 672,000 fiber liners being produced with the new suspension system.

The Hawley Products Company manufactured 3,977,000 fiber liners for McCord. Of the total liners, only three percent were sub-contracted to a second company, the General Fiber Company of St. Louis, Missouri. Production of the fiber liners did not begin until late 1941. The fiber liner continued in production until mid-November 1942, at which time production of the fiber liner was discontinued.

Interior view of an M-1 high-pressure plastic helmet liner, ca. early 1943-June 1944. This particular helmet was manufactured by the Seaman Paper Company.

M-1 helmet liner adjustable head band, ca. 1942-1944. This variation of the band was made of triple Herringbone Twill cotton in olive shade no. 3 with a steel stamped buckle painted olive drab.

M-1 helmet liner chin straps. The foreground chin strap was made of steel painted olive drab and had a rolled edge, ca. 1943-1944. The background chin strap was made of steel painted olive drab and had a straight edge, ca. 1942-1943.

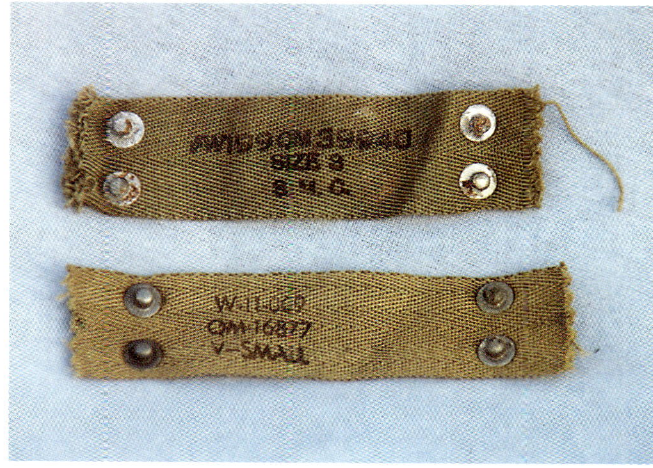

M-1 helmet liner neck bands. These variations of the neck bands were made of single Herringbone Twill cotton in olive shade no. 3 with steel male snap fasteners. The neck band on top was stamped with a size number, ca. 1942-1944. The neck band on the bottom was stamped with the word small, ca. 1944-1945.

M-1 PLASTIC LINER DEVELOPMENT

The fiber liner was considered unsatisfactory almost from its initial production, but considering there were no other alternatives, it was the best possible liner available at the time if the M-1 was to become a standard issue item of the U.S. forces.

The fiber liner had been in production for just a few months and a number of problems in its performance began to surface. The liner most notably tended to absorb a great deal of water and thus lost most of its strength and durability. This would have grave consequences when the liner was issued to troops in the jungle, where the liner would literally become soggy and be of no use. Another problem which faced the liner was that it rather quickly lost its "neat" appearance due to the fact that the cloth outer layer was easily frayed and soiled. With these problems continuing to plague the liner, the U.S. Army sought to continue research on the plastic helmet liner.

In July 1941, research on the plastic helmet liner had begun. By August 1941, the Quartermaster sought to bring together industrial companies, which were familiar with the production of plastic items, to consider the possible replacement of the fiber liner by a laminated plastic liner. The industrial companies participating in the development of the plastic liner were soon given a small contract and asked to produce sample plastic liners, which they felt would meet the current need.

Exterior front right view of an M-1 high-pressure plastic helmet liner, ca. June 1944-August 1945. This particular helmet was manufactured by the Firestone Tire & Rubber Company.

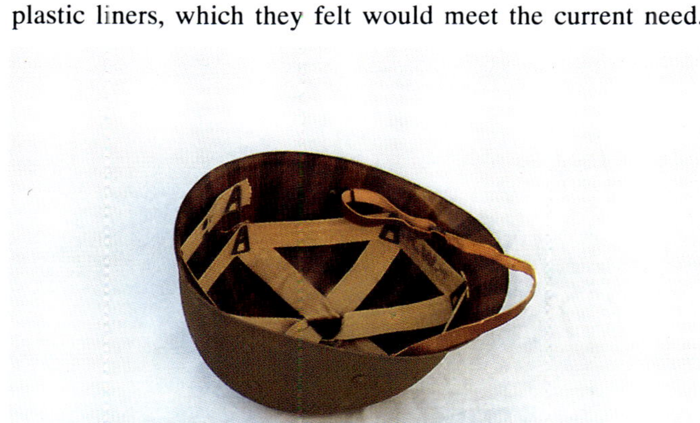

Interior view of an M-1 high-pressure plastic helmet liner, ca. June 1944-August 1945. This particular helmet was manufactured by the Firestone Tire & Rubber Company.

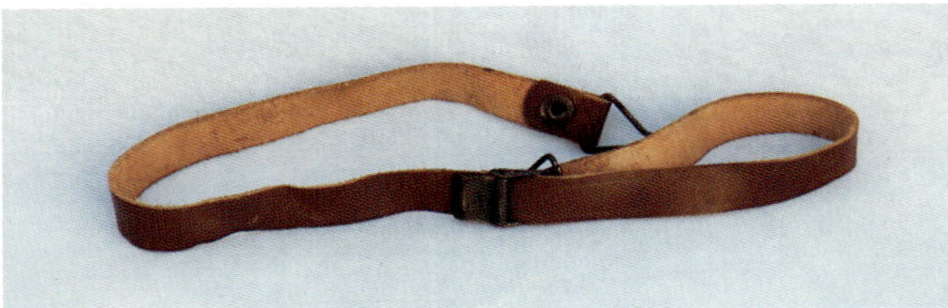

M-1 helmet liner chin strap, ca. 1944-1945. The chin strap was made of brass coated with a mildew inhibitor and had a rolled edge.

M-1 helmet liner adjustable head band, ca. 1944-1945. This variation of the band was made of single Herringbone Twill cotton in olive shade no. 3 with a brass stamped buckle coated with a mildew inhibitor.

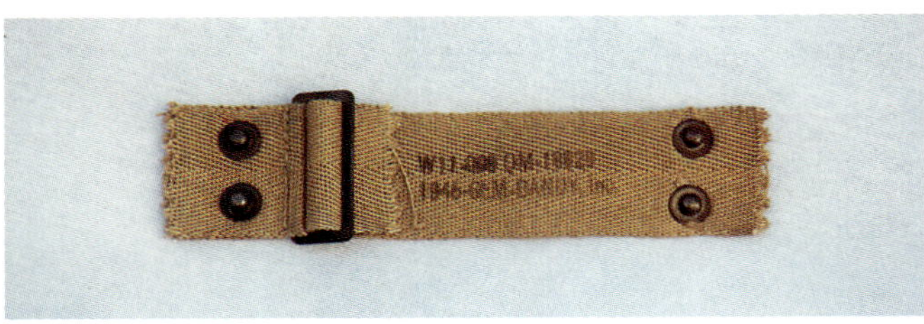

M-1 helmet liner adjustable neck band, ca. 1945. This variation of the band was made of single Herringbone Twill cotton in olive shade no. 3 with brass male snap fasteners and stamped buckle.

M-1 plastic helmet liner crown stamp, mold-in marking, ca. 1942-1945. This variation of the mold-in marking shows the Mine Safety Appliances stamp.

During this time of development, the U.S. Army had not determined, to any great extent, the specification of the liner body and as such it was up to the individual companies to use their expertise in the field of plastics to come up with the best possible plastic liner body. During the fall the various industrial companies submitted their plastic liners for evaluation.

In November 1941, the Office of the Quartermaster General Standardization Branch and the National Bureau of Standards had concluded that the best plastic helmet liner was one made of phenol formaldehyde-impregnated cloth laminated by high-pressure. This conclusion was based on tests the Army conducted on the sample helmet liners submitted during the fall. The tests allowed the Army to establish the durability of each type of sample liner.

After 7 December 1941, the Office of the Quartermaster General requested that a lot of 600 sample liners be produced for further testing. All of these liners were to be produced on sample high-pressure molds developed by the Inland Manufacturing Division of General Motors, and a third of each were to be mixed according to the formula developed by Inland Manufacturing, Mine Safety Appliances, and Westinghouse Electric to determine which of the three provided the best properties.

The results of the high-pressure molding tests were available by late January 1942. The need for a plastic liner was clearly seen in the test report as it indicated that the plastic liners were far superior to the fiber liner. Thus, on 3 February 1942, the Quartermaster Corps adopted the Westinghouse formula type, cloth impregnated helmet liner body. By 11 February 1942 a tentative specification had been written.

M-1 PLASTIC LINER PRODUCTION, COTTON, 1942 - 1945

Contracts for the first plastic helmet liners were made by late February 1942. The initial production of the plastic helmet liner was contracted out to four companies and they were the Inland

Exterior right front view of an M-1plastic helmet liner, ca. 1951-1954. This particular helmet was manufactured by the Westinghouse Electric Company in 1952.

Interior view of an M-1plastic helmet liner, ca. 1951-1954. This particular helmet was manufactured by the Westinghouse Electric Company in 1952.

Manufacturing Division of General Motors, Dayton, Ohio; Micarta Division of the Westinghouse Electric and Manufacturing Company, Trafford, Pennsylvania; the Mine Safety Appliances Company, Pittsburgh, Pennsylvania; and the St. Clair Rubber Company, Marysville, Michigan. The St. Clair Company, had joined the development program late, and had also submitted their low-pressure helmet to the test trials. While their helmet liners did not perform as well as the high-pressure liners, their method of low-pressure manufacture, instead of high-pressure, allowed for a greater number of plastic helmet liners to be more rapidly produced.

By April 1942, the second and last group of companies to receive plastic helmet liner contracts were chosen and they were the Capac Manufacturing Company, Capac, Michigan; Firestone Tire & Rubber Company, Akron, Ohio; Hood Rubber Company, Inc., Watertown, Massachusetts; International Molded Plastics, Inc., Cleveland, Ohio; and the Seaman Paper Company, Chicago, Illinois. All of these new manufacturers adopted the high-pressure style of production with the exception of the Hood Rubber Company, which adopted its own manufacturing process for the helmet, known as the ball-winding low-pressure method. The ball-winding low-pressure method was allowed because it offered a less expensive form of production.

The plastic helmet liner was manufactured by the following method. Cotton ducking material was first impregnated with resin and then cut into various shaped segments. These shaped segments were then arranged and stapled or cemented together to create a pre-formed helmet liner. An extra piece of impregnated duck material was added to the interior crown for strength. The pre-formed helmet was then placed into a mold. For the high-pressure liners the male and female molds were made of steel and high-pressure force was applied and the helmet was then allowed to cure. For the low-pressure liners the male mold was rubber and female molds were made of steel and low-pressure force was applied and the helmet was then allowed to cure. The result was a hard, shaped helmet body. This hard, shaped helmet body had a smooth finish both inside and out if made of high pressure, and only a smooth exterior if made of low pressure. The next operation called for the removal of excess flash from around the helmet liner edge, this was pre-formed by a punch press. Next the edges of the helmet were sealed by burnishing. This helmet body was then finished by first being pierced by a hole punch, the holes being needed for the attachment by riveting of the suspension, neck strap, insignia eyelet, and the studs for the chin strap, and then secondly by the riveting of these items. The final process called for the helmet to be painted and baked dried. The paint was only applied to the exterior of the liner and was olive drab in color. The exception being the early St. Clair liners which had interior paint. In some plants, painting was accomplished using automatic paint spray machines, while other plants used hand paint spray. Baking of the paint was performed by either using an oven of infra-red lamps for two minutes, or using a horizontal drying ovens for fifteen minutes.

The finished liner varied in thickness but averaged about 0.082" thick, and measured 8.6" wide by 10.6" long. The weight of the liner with all of its components assembled was 0.75 pounds. The helmet liner shell, when finished, also contained a small molded marking on the interior of the crown to indicate the manufacturing company. The interior, which was not painted, possessed a brown striped design which was attributed to the impregnated duck fabric segments used in the manufacturing method.

PLASTIC HELMET LINER COMPONENTS AND PRODUCTION CHANGES, 1942 - 1945

The plastic helmet liner initially produced incorporated the Riddell suspension system with the rayon webbing as used in the earlier Hawley fiber liner. Approximately 3,600,000 plastic helmet liners were manufactured in this way before production was changed to use the new cotton suspension system in the summer of 1942.

The new cotton suspension system contained the following changes. The rayon webbing was to be replaced by a cotton webbing of khaki color, olive drab shade no. 3, which was manufactured in a single Herringbone twill weave. The single Herringbone twill weave produced a v-type pattern in the webbing. To secure this new webbing to the liner a new washer was developed. This new, Westinghouse designed, A-shaped washer replaced the older perpendicular sided washers. The hammock style suspension still retained a shoelace for adjustment as it did when the suspension was made of rayon. The basic Riddell web-type suspension remained but many of its smaller components were changed.

A new adjustable head band was developed, and was made adjustable by use of a two piece bar buckle. The new head band was also designed to have vegetable tanned calfskin sewn on one side for its entire length, except at its ends, to allow adjustment by the bar buckle. A new chin strap buckle for the liner and a new spring clip used to attach the head band to the suspension were developed. The new buckle replaced the older square two slit buckle, which was considered too hard to adjust by the Army. Adjustment was made possible by use of a cam lever, which had an upwardly curled rear resilient section.

In response to the Army's desire for a new method of attaching the leather chin strap to the liner, the United Carr Fastener Company developed a new fastener. The fastener consisted of a garter stud and hook that permitted the chin strap to be securely fastened to the liner and yet removable.

All of these items became standard for production in summer of 1942, and all the hardware items were to be made of steel painted with a coat of olive drab. Stamp markings often appeared on the new items and webbing. The head band and neck band usually displayed the contract number and the manufacturer for those items. The neck band also included a size number. The introduction of these new items not only affected the plastic helmet liner production but were also used on the remaining fiber liners which did not complete production until November 1942.

In the spring of 1943, two refinements were made to the helmet liner, which invloved the paint and the bar buckle. A suitable finish had been developed for the plastic helmet liner. The original paint mixture produced plastic helmet liners with a dark olive drab color and a smooth finish. Between October 1942 and March 1943, all of the plastic helmet liner manufacturers had switched to the new paint mixture which contained small particles of phenolic resin and produced a textured finish . The new paint appeared to be a lighter shade of olive drab. A new two piece bar buckle debuted. The new buckle consisted of a single piece of metal, which was similar to a rectangular two slit buckle, except that one slit had rounded teeth on one side, while the other slit was completely smooth.

In the fall of 1943, the chin strap buckle clip design was also modified from the straight edge to a roller, to allow any thickness of leather chin strap to be admitted.

By June 1944, brass was adopted for use in the manufacture of all hardware items for the helmet liner, with the exception of the spring clips. The brass used in the hardware of the liner was coated with a rust and mildew inhibitor that gave the brass items a flat black coat like finish. Other changes in the specification during the spring of 1944, allowed for the use of double and triple Herringbone twill weave, instead of the single Herringbone twill weave, in the production of the khaki, olive drab shade no. 3, cotton webbing. The specification also called for the neck band's size number to be replaced by a stamp with the word small, medium, or large.

M-1 helmet liner head band, ca. 1951-1954. The head band was made of single Herringbone Twill cotton in olive drab shade no.7 with brass buckle painted light olive drab. Note specification number, MIL-B-1853.

M-1 helmet liner neck band, ca. 1951-1954. The neck band was made of single Herringbone Twill cotton in olive drab shade no.7 with a brass buckle painted light olive drab, and brass male snap fasteners. Note specification number, MIL-B-1853.

M-1 helmet liner chin strap, ca. 1951-1954. The chin strap buckle and holders were made of steel painted black. Note the slight curve of the buckle lever.

M-1 plastic helmet liner crown stamp, mold-in marking, ca. 1951-1954. This variation of the mold-in marking shows the Westinghouse stamp.

In September 1944, a suggestion was made that the sized neck band be replaced by a single adjustable neck band that would fit all sizes. Shortly thereafter, action was taken and a single adjustable neck band was manufactured and issued.

PLASTIC HELMET LINER TOTALS, 1941 - 1945

Ten companies produced the helmet liner shell, while an additional 30 were responsible for the manufacture of the various components that went into the liner. The total number of plastic liners produced, were at least 39,723,000 during the years between 1941-1945. Helmet liner production was discontinued around 17 August 1945. Westinghouse was the largest manufacturer of the helmet liner, accounting for about half of the liners produced, with its Micarta Division producing approximately 13,000,000 and its Bryant Electric subsidiary producing approximately 10,000,000. Firestone Tire & Rubber produced approximately 7,500,000 plastic liners, and the Inland Manufacturing Division of General Motors completed approximately 1,900,000 liners. The remaining companies, Mine Safety Appliances, Capac, Seaman, and International Molded Plastics each produced between 2,000,000 and 4,000,000 plastic helmet liners. Production of the St. Clair plastic helmet liners remained low with only 1,300,000 being produced between spring 1942 and 1943. Production by the Hood Rubber Company also remained low with only 206,000 plastic helmet liners being manufactured in 1943.

PLASTIC LINER PRODUCTION, COTTON, 1950 - 1954

In 1951, production of the M-1 helmet resumed. These production liners looked similar to the final production World War

M-1 plastic helmet liner interior stamp patch, ca. 1951-1954. This variation of the stamp patch shows the Westinghouse logo, the division name and year of material manufacture, in this case, Micarta Division, 1952.

U.S. Army solider wearing the M-1 helmet liner, ca. 1951-1954. The soldier is Joseph G. Reynosa.

II plastic liner, but differed by incorporating several additional changes. All of the liners interior webbing, the suspension, neck strap, head band and neck band, were manufactured using olive drab shade no. 7 cotton single Herringbone Twill. The leather liner chin strap also had manufacturing changes. The liner chin strap buckle and the chin strap holders were manufactured of steel and painted black. The liner retained its eyelet hole.

During this period the Micarta Division of Westinghouse Electric Company, became the sole supplier of impregnated cotton duck for the production of plastic helmet liners. The number of manufacturers producing plastic helmet liners between 1951 and 1954 was three and included the Micarta Division of Westinghouse Electric Company, Capac Manufacturing Company, and the Mine Safety Appliances Company. A manufacturing stamp appeared as a molded stamp on the interior crown. The molded stamp on Capac helmet liners included the word CAPAC itself and a Westinghouse molded stamp to acknowledge Westinghouse's supply of cotton duck. The manufacturing stamps on the Westinghouse and Mine Safety Appliances helmet liners included their respective molded stamps and a Westinghouse ink stamp on a square piece of impregnated material to acknowledge Westinghouse's supply of cotton duck. The Westinghouse ink stamp included a Westinghouse logo and the date of manufacture for the cotton duck.

PLASTIC LINER PRODUCTION, COTTON, 1955 - 1962

In 1955, a number of changes were instituted in the production liners as well as changes in in manufacturers. By 1955 the Mine Safety Appliances Company appeared to have dropped out of liner production. Also in that year the Firestone Tire & Rubber Company appeared to have rejoined the production effort. While Westinghouse continued to supply cotton duck for itself and Capac, Firestone obtained cotton duck from other sources.

Changes for the liner included the change to olive drab shade no. 7 cotton "heavy" webbing in place of olive drab shade no. 7 cotton Herringbone Twill webbing. The insignia eyelet hole was dropped from production. The leather liner chin strap wedge buckle adopted a design change that included an acute angle for the cam lever, which previoulsly had an upwardly curled rear resilient section. The leather liner chin strap wedge buckle and chin strap holders continued to be made of steel painted black.

A manufacturing stamp appeared as a molded stamp on the interior crown. The molded stamp on Capac helmet liners included the Capac and Westinghouse molded stamp, and the Westinghouse ink stamp. The manufacturing stamps on the

Exterior right front view of an M-1 plastic helmet liner, ca. 1955-1962. This particular helmet was manufactured by the Westinghouse Electric Company in 1955.

Interior view of an M-1 plastic helmet liner, ca. 1955-1962. This particular helmet was manufactured by the Westinghouse Electric Company in 1955.

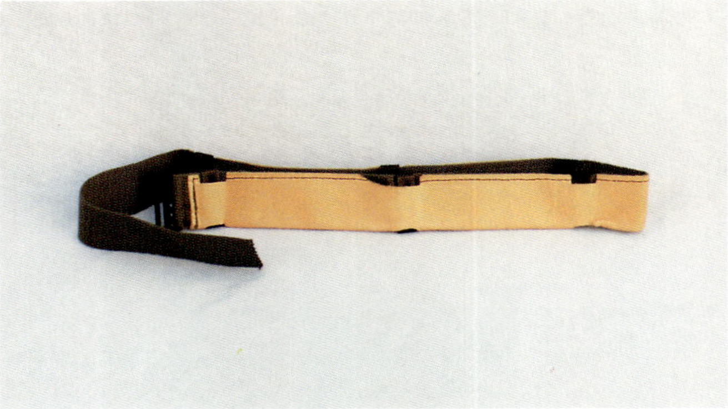

M-1 helmet liner head band, ca. 1955-1962. This variation head band was made of cotton in olive drab shade no.7 with a steel buckle painted black. Note unfinished webbing end.

Westinghouse helmet liners included Westinghouse molded stamp and an Westinghouse ink stamp. In 1961, the Westinghouse ink stamp changed to include the new Westinghouse logo and the date of manufacture for the cotton duck. The molded stamp on Firestone helmet liners included only the stylized F without the shield. Capac ink stamps also appear during this period.

PLASTIC LINER PRODUCTION, COTTON, 1963 - 1968

In 1963, the U.S. Army introduced a new suspension system for the M-1 helmet liner. The new suspension consisted of a head suspension band and cradle. The cradle was composed of three adjustable webbing cradle straps. Adjustment of the cradle straps was made possible by use of stamped bar buckles. The

M-1 helmet liner neck band, ca. 1955-1962. The neck band was made of cotton in olive drab shade no.7 with a steel buckle painted black, and brass male snap fasteners. (Courtesy of Larry Sutherland)

M-1 helmet liner chin strap, ca. 1955-1962. The chin strap buckle and holders were made of steel painted black. Note the acute angle at the end of the buckle lever.

M-1 plastic helmet liner crown stamp, mold-in marking, ca. 1955-1962. This variation of the mold-in marking shows the Westinghouse stamp.

M-1 plastic helmet liner interior stamp patch, ca. 1955-1962. This variation of the stamp patch shows the Westinghouse logo, the division name and year of material manufacture, in this case, Micarta Division, 1955. Note the webbing texture.

Exterior left front view of an M-1 plastic helmet liner, cotton, ca. 1963-1968. This particular helmet was manufactured by the Firestone Tire & Rubber Company in 1965.

bar buckles were made of steel and painted black. Stitched to the center webbing cradle strap was a label. The liner also included a new neck band assembly, which consisted of three 0.5" wide webbing straps attached to a cotton textile webbed neck band body. The three neck band webbing straps were secured to the liner by means of three neck band buckle and clip assemblies made of steel and painted black. All of the cotton webbing was olive drab shade no. 7.

The leather liner chin strap was completely dropped from the helmet design in 1963. The U.S. Army did require crown markings for liners produced between 1963 and 1968. This crown marking could either be a molded crown mark or take some other form, i.e. ink or paint. Some production liners were manufactured with no molded crown markings, while other helmets were produced with either the Firestone stylized F or an unidentified manufacturers U S molded crown marking. During this period of production the U.S. Army did require a printed identification marking in ink centered on the inside surface of the head suspension band. At a minimum the printed legend was to contain the following:

LINER, SOLIDER'S STEEL HELMET, M-1
Date and number of contract

Interior view of an M-1 plastic helmet liner, cotton, ca. 1963-1968. This particular helmet was manufactured by the Firestone Tire & Rubber Company in 1965.

M-1 helmet liner suspension stamp marking, ca. 1963-1968. This view shows the stamp marking which reads "LINER, SOLDIER'S STEEL HELMET, M-1 TYPE 1". This particular helmet was manufactured in 1965.

Exterior front left view of an M-1 plastic helmet liner, nylon, ca. 1963-1968.

Interior view of an M-1 plastic helmet liner, nylon, ca. 1969-1988. This particular helmet was manufactured by Specialty Plastic Products. Note the detachable suspension.

Exterior front left view of an M-1 plastic helmet liner, nylon, ca. 1969-1988. This particular helmet was manufactured by Specialty Plastic Products.

Interior view of an M-1 plastic helmet liner, nylon, ca. 1963-1968.

M-1 nylon helmet liner, ca. 1963 (Courtesy of U.S. Army)

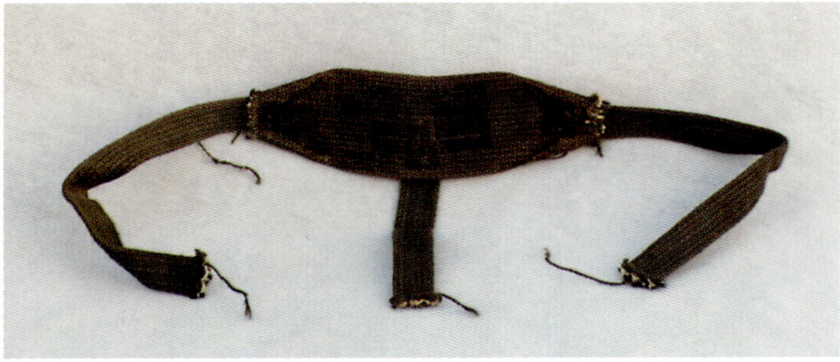

M-1 helmet liner neck strap, ca. 1963-1988. The neck strap was made of cotton in olive drab shade no.7.

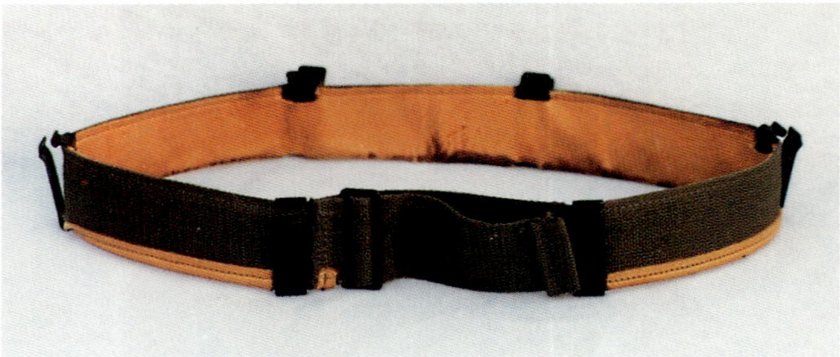

M-1 helmet liner head band, ca. 1963-1988. This variation head band was made of cotton in olive drab shade no.7 with a steel buckle painted black. Note finished webbing end.

Interior view of an M-1 plastic helmet liner, nylon, without suspension, ca. 1969-1988. This particular helmet was manufactured by Specialty Plastic Products.

M-1 helmet liner detachable suspension, ca. 1969-1988.

Firestone and another unidentified company appear to have been the only manufacturers of the cotton plastic helmet liners between 1963 and 1968.

PLASTIC LINER PRODUCTION, COTTON, 1969 - 1975

In 1969, U.S. specifications indicated that the cotton plastic helmet liner was to be manufactured with a new removeable suspension. Additonally the specification, once again, required

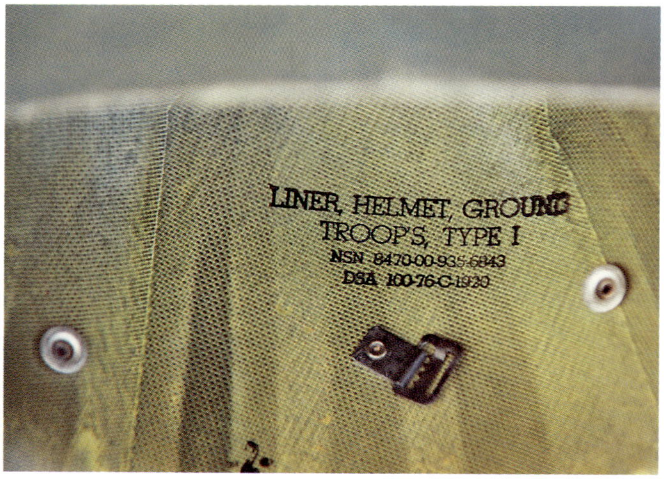

M-1 helmet liner, nylon, identification marking, ca. 1969-1988. The marking reads "LINER, HELMET, GROUND TROOPS, TYPE I". Below this is the National Stock Number, NSN, and the DSA.

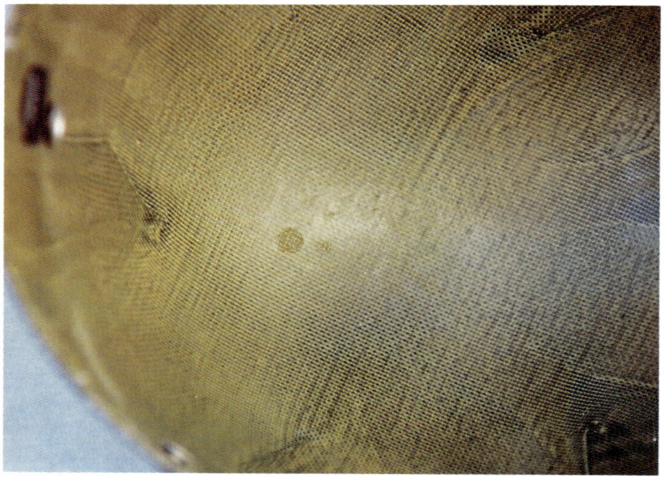

M-1 helmet liner, nylon, stamp marking, ca. 1969-1988. The stamp marking, mold-in mark, show the marking for Special Plastic Products.

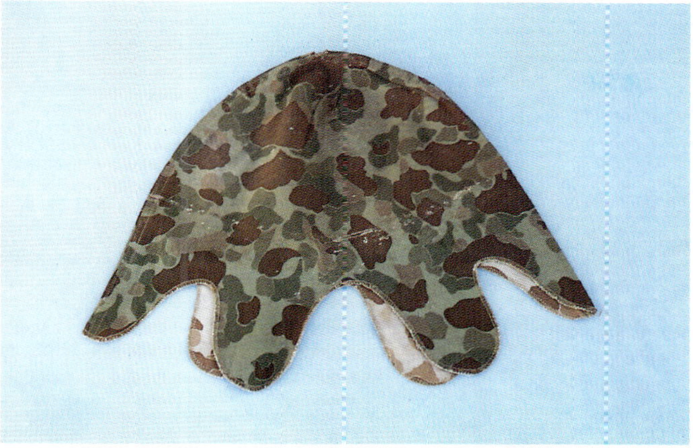

M-1 USMC camouflage helmet cover, "greenside" side out, September 1942-August 1945.

M-1 USMC camouflage helmet cover, "brownside" side out, September 1942-August 1945.

the use of a molded crown marking. However, it appeared that no cotton plastic helmet liners were manufactured to this specification and the cotton plastic helmet liner was cancelled shortly thereafter. In 1975 the specification for the cotton plastic helmet liner was cancelled.

PLASTIC LINER PRODUCTION, NYLON, 1963 - 1968

In 1962, the U.S. Army developed a new impregnating material for the helmet liner, laminated nylon. The laminated nylon improved the ballistic qualities of the liner over the impregnated cotton. The nylon fabric was impreganted with phenolic-butyral on the inner and straight phenolic on the outer surface of the liner. By 1963, the nylon fabric liners were in production. The nylon fabric resulted in a number of interior combinational strippings that ranged in color from various light green shades to yellow, or orange.

The nlyon liners were produced with the new suspension system introduced on the cotton fabric liners in 1963. The new suspension consisted of a head suspension band and cradle. The cradle was composed of three adjustable webbing cradle straps. Adjustment of the cradle straps was made possible by use of stamped bar buckles. The bar buckles were made of steel and painted black. Stitched to the center webbing cradle strap was a label. The liner also included a new neck band assembly, which consisted of three 0.5" wide webbing straps attached to a cotton textile webbed neck band body. The three

Front right view of M-1 helmet with USMC camouflage helmet cover, "brownside" out, ca. September 1942-August 1945.

M-1 USMC camouflage helmet cover with buttonholes, "greenside" side out, September 1942-August 1945.

Front left view of M-1 helmet with USMC camouflage helmet cover with buttonholes, "greenside" out, ca. September 1942-August 1945.

M-1 USMC camouflage helmet cover with buttonholes, "greenside" side out, ca. 1953-1954. This particular cover was manufactured in 1953.

neck band webbing straps were secured to the liner by means of three neck band buckle and clip assemblies made of steel and painted black. All of the cotton webbing was olive drab shade no. 7. The leather liner chin strap was completely dropped from the helmet design in 1963.

As with the cotton plastic helmet liners, the U.S. Army required markings on the helmet liners. The U.S. Army did require crown markings for liners produced between 1963 and 1968. This crown marking could either be a molded crown mark or take some other form, i.e. ink or paint. Some production liners were manufactured with no molded crown markings, while other helmets were produced with either the Firestone stylized F or an unidentified manufacture's U S molded crown marking. During this period of production the U.S. Army did require a printed identification marking in ink

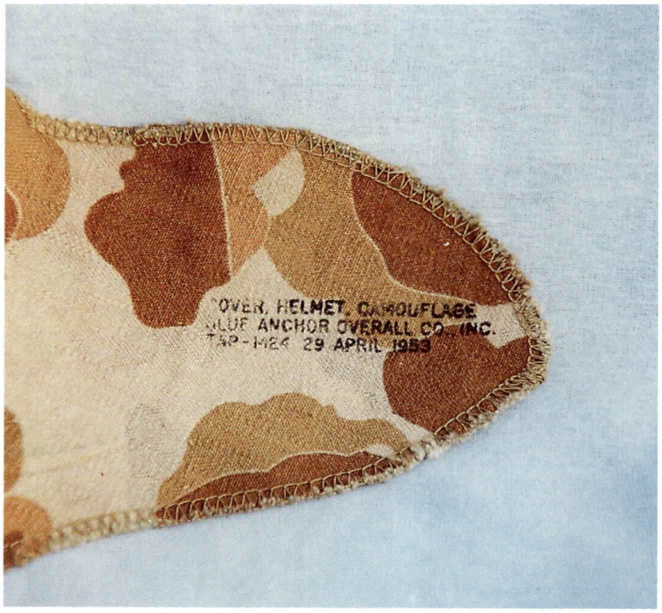

Close-up of stamp marking on USMC helmet cover, ca. 1953-1954. This particular cover was manufactured in 1953.

Front right view of M-1 helmet with USMC camouflage helmet cover with buttonholes, "brownside" out, ca. 1953-1954.

M-1 camouflage helmet cover with buttonholes, leaf pattern, "greenside" side out, ca. 1960-1965. The cover was made of cotton twill material.

M-1 camouflage helmet cover with buttonholes, leaf pattern, "brownside" side out, ca. 1960-1965. The cover was made of cotton twill material.

centered on the inside surface of the head suspension band. At a minimum the printed legend was to contain the following:

> *LINER, SOLIDER'S STEEL HELMET, M-1*
> *Date and number of contract*

Firestone and another unidentified company appear to have been the only manufacturers of the nylon plastic helmet liners between 1963 and 1968.

PLASTIC LINER PRODUCTION, NYLON, 1969 - 1988

In 1969, U.S. specifications called for the nylon plastic helmet liner was to be manufactured with a new removable suspension. Additionally, the specification, once again, required the use of a molded crown marking. During this period of production the U.S. Army did require a printed identification marking in ink centered on the inside surface of the liner. Initially the identification marking was printed to an adhesive tag, later it was printed directly to the liner's interior. At a minimum the printed legend was to contain the following:

Close-up of Eagle, Globe and Anchor stamp on USMC helmet cover, ca. 1953-1954. Note double riband in eagle's beak, and rope on anchor.

Close-up of stamp marking on M-1 helmet cover, ca. 1960-1965.

Front left view of M-1 helmet with camouflage helmet cover with buttonholes, "greenside" out, ca. 1960-1965.

LINER, GROUND TROOP'S HELMET, M-1
Date and number of contract

Specialty Plastic Products and Firestone Tire and Rubber Company appear to have been the only manufacturer of the nylon plastic helmet liners between 1969 and 1988. The specification for the nylon plastic helmet liner continued through 1988.

M-1 camouflage helmet cover with buttonholes, leaf pattern, "greenside" side out, ca. 1966-1970. The cover was made of cotton duck material. Note deepened gap between two middle flaps to provide clearance for helmet body chin strap.

M-1 camouflage helmet cover with buttonholes, leaf pattern, "greenside" side out, ca. 1966-1970. The cover was made of cotton duck material.

CAMOUFLAGE HELMET COVERS, 1942 - 1945

In September 1942, the United States Marine Corps adopted the camouflaged helmet cover for use with the M-1 Helmet. The probable use of the helmet cover was to provide the wearer with additional camouflage properties in at least two different types of environments. The cover was manufactured in two different color combinations, both using the same camouflage pattern. The colors of the helmet cover suggest that the cover

Side view of M-1 helmet with camouflage helmet cover with buttonholes, "greenside" out, ca. 1966-1970.

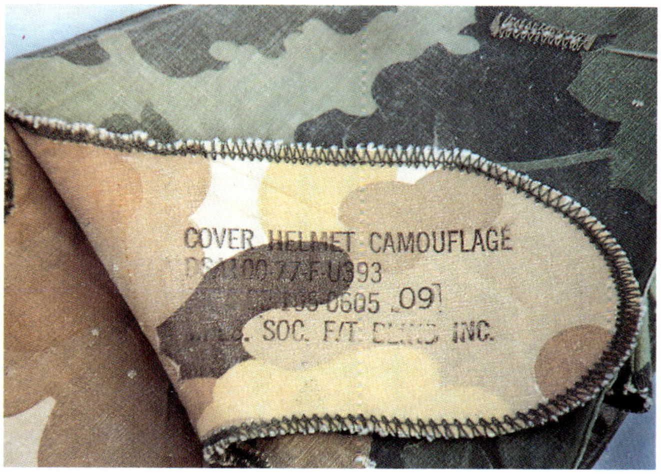

Close-up of stamp marking on M-1 helmet cover, ca. 1966-1970.

Side view of M-1 helmet with camouflage helmet cover, "leaf" pattern, "brownside" out, ca. 1966-1970.

Front right view of M-1 helmet complete, ca. 1966-1970.

Rear left view of M-1 helmet complete, ca. 1966-1970.

U.S. Army soldier wearing the M-1 helmet with "leaf" cover, "brownside" out, and camouflage band. Note the helmet body chin strap clip length, which indicates the helmet was circa 1966-1972. Note also the cover cut around the chin strap loop, which indicates that this was the cotton "leaf" cover. (Courtesy of U.S. Army)

U.S. Army dog handler, Carlos Vidal, wearing the M-1 helmet with "leaf" pattern camouflage helmet cover, Vietnam, ca. 1970. (Courtesy of Carlos Vidal)

U.S. Army soldier wearing the M-1 helmet with "leaf" cover, "greenside" out, and camouflage band. Note the helmet body chin strap clip length, which indicates the helmet was circa 1961-1965. Note also the cover cut around the chin strap loop, which indicates that this was the cotton "leaf" cover. (Courtesy of U.S. Army)

Members of the USAF wear their M-1 helmets, camouflage covers and camouflage bands at Pleiku A.B., December 1968. From left to right: A1C Sanders wears M-1 helmet with "leaf" pattern cover, "greenside" out; Sgt. Hinson wears M-1 helmet with "leaf" pattern cover, "greenside" out and camouflage band; Sgt. Wicks wears M-1 helmet with "leaf" pattern cover, "brownside" out and camouflage band; A1C Sutherland wears M-1 helmet with "leaf" pattern cover, "brownside" out; and Sgt. Wormsley wears M-1 helmet with "leaf" pattern cover, "greenside" out and camouflage band. (Courtesy of Larry Sutherland)

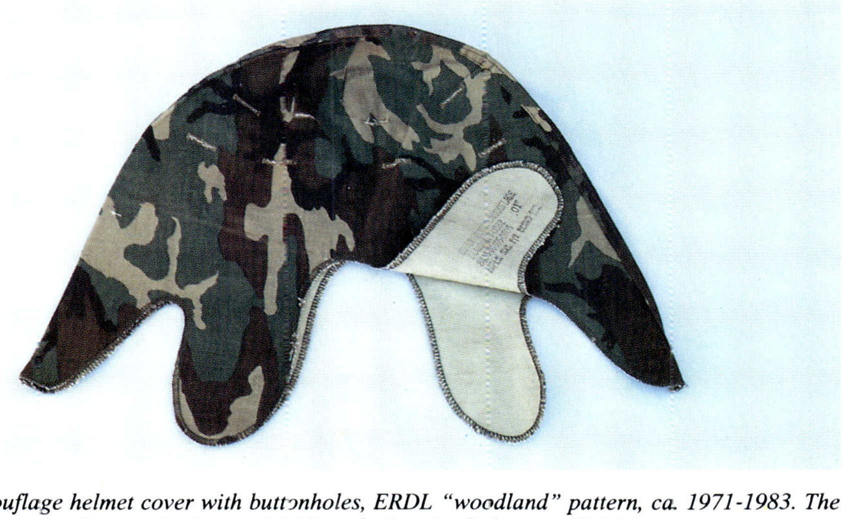

M-1 camouflage helmet cover with buttonholes, ERDL "woodland" pattern, ca. 1971-1983. The cover was made of cotton duck material. Note the interior light green finish.

Side view of M-1 helmet with camouflage helmet cover, ERDL "woodland" pattern, ca. 1971-1983.

was to be used in either a jungle type environment or a beach/desert environment.

The cover was manufactured of cotton Herringbone twill cloth, and consisted of a "greenside" and a "brownside". The "greenside" consisted of a light green background with a four color pattern of medium and light, brown and green. The "brownside" consisted of a tan background with a three color pattern of medium, light, and very light brown. The cover was manufactured of two identical halves, which when sewn together created a dome-like piece of material that fitted neatly over the helmet body. The cover included flaps which were sandwiched between the helmet liner and the helmet body when the cover was used, or which could be left out to cover the neck or ears if the wearer chose to break up the outline of the helmet. The cover was produced with and without buttonholes. These buttonholes provided a means of fastening additional camouflage material, such as surrounding vegetation.

The Marine Corps camouflage helmet covers were manufactured between 1942 and 1945. It is known that the cover was only issued to Marine Corps troops during World War II.

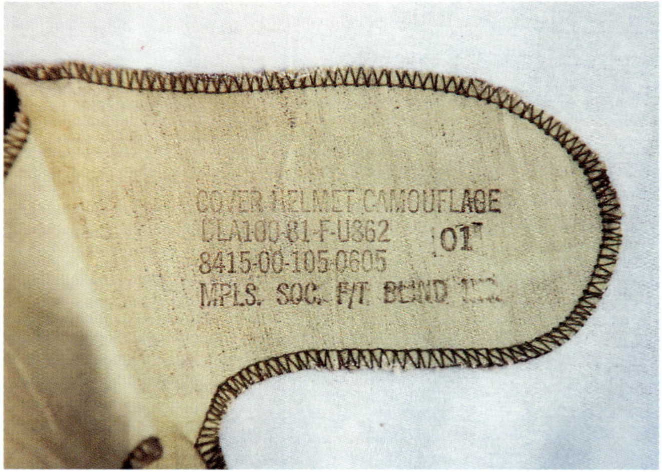

Close-up of stamp marking on M-1 helmet cover, ca. 1971-1983.

M-1 camoufalge helmet cover with buttonholes, ERDL "woodland" pattern, ca. 1971-1983. The cover was made of cotton duck material. Note the unfinished interior.

U.S. Army soldier in Europe wearing the M-1 helmet with early ERDL "woodland" cover and camouflage band. (Courtesy of U.S. Army)

U.S. Army soldiers wearing the M-1 helmet complete with the ERDL "woodland" cover during operation Urgent Fury in Grenada, 1983. (Courtesy of U.S. Army)

M-1 camoufalge helmet cover with buttonholes, "woodland" pattern, ca. 1983-1989. The cover was made of cotton/nylon material. Note the interior grey finish.

Side view of M-1 helmet with camouflage helmet cover, "woodland" pattern, ca. 1983-1989.

CAMOUFLAGE HELMET COVERS, 1953 - 1954

During 1953 and 1954, production of the Marine Corps camouflage helmet covers resumed. The cover remained the same as the earlier manufacture with some revisions. The major revision saw the single buttonhole for each flap no longer warranted for production. Additionally, the covers were stamped with the item name, the manufacturer, and the contract date on one of the flaps. The Blue Anchor Overall Company appeared to be the sole manufacturer of the Marine Corps camouflage helmet cover between 1953 and 1954.

CAMOUFLAGE HELMET COVERS, 1960 - 1965

In 1959, the Marine Corps had developed a new camouflage helmet cover for the M-1 helmet. The new cover was manufactured of cotton Herringbone twill cloth type material and

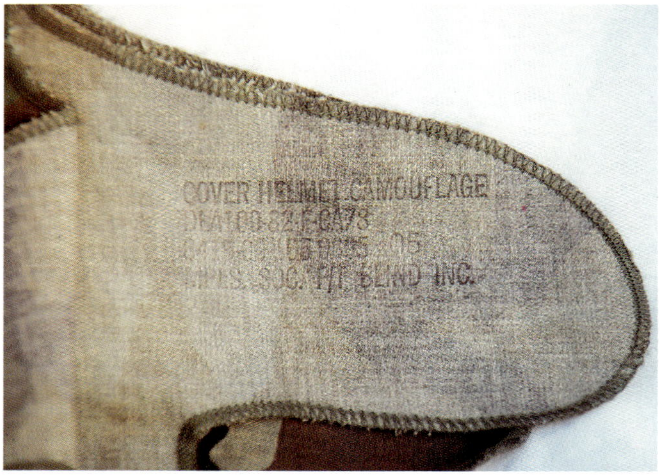

Close-up of stamp marking on M-1 helmet cover, ca. 1983-1989.

Camouflage helmet band, ca. 1950-1981.

Close-up of stamp marking on camouflage helmet band, ca. 1950-1981.

Front left view of M-1 helmet complete, ca. 1983-1989.

Rear right view of M-1 helmet complete, ca. 1983-1989.

was reversible, consisting of a "leaf" patterns in green colors and brown colors.

The "greenside" consisted of a light green background with a five color pattern of light green, medium green, dark green, buff, and dark brown. The "brownside" consisted of a very

Camouflage helmet band, ca. 1981-1988. Note luminous tapes.

Close-up of stamp marking on camouflage helmet band, ca. 1981-1989.

Close-up of box stitch on camouflage helmet band, ca. 1950-1981.

Close-up of box stitch and luminous tapes on camouflage helmet band, ca. 1981-1989.

light brown background with a four color pattern of light brown, medium brown, dark brown, and tan.

The cover was produced with buttonholes. These buttonholes provided a means of fastening additional camouflage material, such as surrounding vegetation. The cover contained sixteen crown buttonholes. Additionally, the covers were stamped with the item name and the contract number on one of the flaps.

CAMOUFLAGE HELMET COVERS, 1966 - 1970

In 1966, production of the "leaf" camouflaged helmet cover changed to use cotton duck cloth. The cotton duck cloth replaced the cotton twill material which was heavier in weight. The cover was reversible, consisting of the earlier "leaf" patterns in green colors and brown colors.

The cover was produced with buttonholes. These buttonholes provided a means of fastening additional camouflage material, such as surrounding vegetation. The cover contained sixteen crown buttonholes. The gap between the two side middle flaps was deepened to allow clearance for the helmet body chin strap. Additionally, the covers were stamped with the item name and the contract number on one of the flaps. Among the firms who produced the cover was Mpls. Soc. F/T Blind Inc.

CAMOUFLAGE HELMET COVERS, 1971 - 1983

In 1971, the U.S. Army Engineering Research and Development Laboratory introduced the "woodland" camouflage pattern for use with the M-1 helmet cover. The helmet cover was constructed of cotton duck cloth, was not reversible, and consisted of the ERDL "woodland" pattern in green colors and brown colors. The interior of the cover was either light green in color or left unfinished.

The cover was produced with buttonholes. These buttonholes provided a means of fastening additional camouflage material, such as surrounding vegetation. The cover contained sixteen crown buttonholes. The gap between the two side middle flaps was deepened even more to allow clearance for the helmet body chin strap. Additionally, the covers were stamped with the item name and the contract number on one of the flaps. Among the firms who produced the cover was Mpls. Soc. F/T Blind Inc.

CAMOUFLAGE HELMET COVERS, 1983 - 1989

In 1983, production of the ERDL "woodland" camouflaged helmet cover changed to use cotton/nylon cloth. The cotton/nylon cloth replaced the cotton duck cloth which was heavier in weight. The helmet cover was constructed of cotton/nylon cloth, was not reversible, and consisted of a modified ERDL "woodland" pattern in green colors and brown colors. The interior of the cover was grey in color.

The cover was produced with buttonholes. These buttonholes provided a means of fastening additional camouflage material, such as surrounding vegetation. The cover contained sixteen crown buttonholes. The gap between the two side middle flaps was deepened to allow clearance for the helmet body chin strap. Additionally, the covers were stamped with the item name and the contract number on one of the flaps. Among the firms who produced the cover were Mpls. Soc. F/T Blind Inc.; and Volunteer Blind Industries.

CAMOUFLAGE HELMET BAND, 1942 - 1945

The Camouflage Helmet Band was a standardized item as early as fall of 1942. The band was elastic and constructed of Neoprene, a synthetic rubber. The elastic strip was formed into a band by joining the ends with a box tack stitch with diagonals. The finished band was 23" in circumference and was green in color, olive drab shade no. 7. The band was to go over the helmet body or the liner, and to hold branches, leaves or other vegetation for camouflage.

The U.S. Army Camouflage Helmet Band was manufactured between the fall of 1942, and the summer of 1945. The band also appeared to have been issued only to U.S. Army troops during World War II.

CAMOUFLAGE HELMET BAND, 1950 - 1981

Between 1950 and 1981, production of the Camouflage Helmet Band continued. Its construction remained basically the same as its World War II forbear. The band was 23" in circumference, green in color, olive drab shade no. 7, and joined at the ends with a box tack stitch with diagonals.

CAMOUFLAGE HELMET BAND, 1981 - 1988

In 1981, production of the Camouflage Helmet Band changed to included two luminous tapes at the rear of the band. The band was 23" in circumference, green in color, olive drab shade no. 7, and joined at the ends with a box tack stitch with diagonals.

4

THE M-2, M-1C AND M-1 PARACHUTIST HELMETS

The concept of a separate helmet for parachutists was considered as early as 1941. For this role, the standard M-1 helmet was initially tested, but it was found unsuitable. The standard M-1 helmet had a tendency to be knocked off when the parachute opened or when the parachutist assumed an inverted position. To overcome these problems, the Research and Development Branch of the Quartermaster Corps, along with other departments and private industry, developed the necessary modifications for the M-1 helmet to make it suitable for use by parachute troops.

Close-up of the M-2 helmet body's fixed D-ring loop, web chin strap, and extension with snap fastener, ca. January 1942-summer 1943. (Courtesy of Michel De Trez)

Exterior view of the M-1 parachutist fiber liner, ca. January 1942-November 1942. (Courtesy of Michel De Trez)

Interior view of the M-1 parachutist fiber liner, ca. January 1942-November 1942. (Courtesy of Michel De Trez)

A World War II U.S. Army parachutist wearing the M-2 parachutist helmet body and M-1 parachustist plastic helmet liner, ca. 1943. (Courtesy of U.S. Army via Larry Sutherland)

Interior view of the parachutist helmet body, ca. summer 1943-October 1943.

Exterior view of the parachutist M-1 plastic helmet liner, ca. 1943. The liner was manufactured by the Inland Manufacturing Division of General Motors.

Close-up of the helmet body's fixed loop, web chin strap, and extension with snap fastener, ca. summer 1943-October 1943. (Courtesy of Michel De Trez)

Interior view of the parachutist M-1 plastic helmet liner, ca. 1943. The liner was manufactured by the Inland Manufacturing Division of General Motors.

Close-up of the helmet body's flexible loop, web chin strap, and extension with snap fastener, ca. October 1943-November 1944. (Courtesy of Michel De Trez)

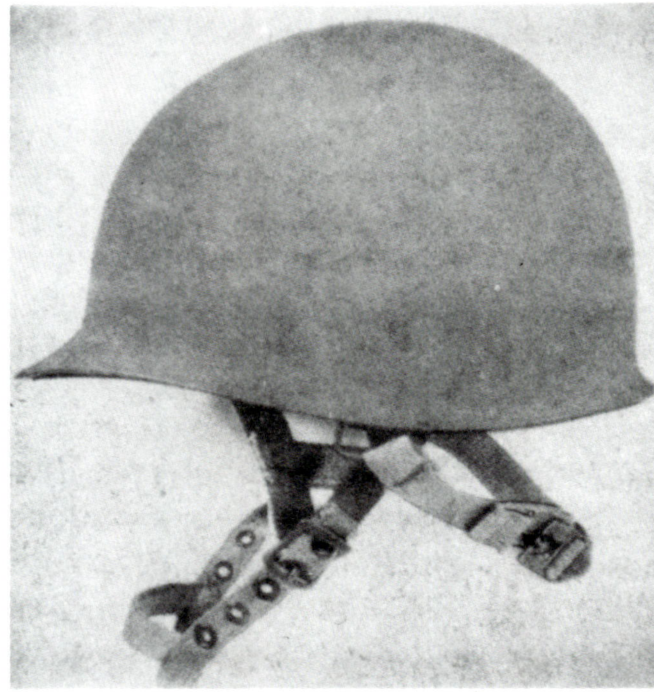

Exterior view of M-1C parachutist helmet, ca. January 1945-August 1945. (Courtesy of Office of the Chief of Ordnance, U.S. Army)

Interior view of the parachutist M-1 plastic helmet liner, ca. late 1944. The liner was manufactured by the Westinghouse Electric Company. Note the olive drab shade no. 7 straps and wire buckle.

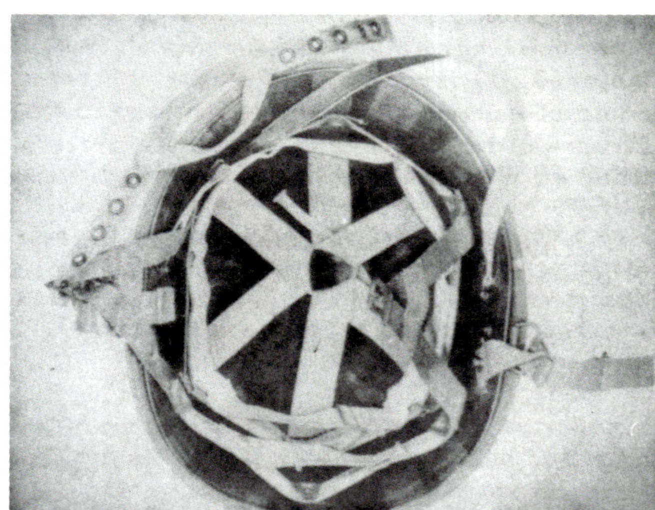

Interior view of M-1C parachutist helmet, ca. January 1945-August 1945. (Courtesy of Office of the Chief of Ordnance, U.S. Army)

Exterior left front view of the M-1C helmet body, ca. January 1945-August 1945.

Interior view of the M-1C helmet body, ca. January 1945-August 1945.

Close-up of the flexible chin strap loop and olive drab shade no. 7 webbed chin strap on the M-1C helmet body, ca. January 1945-August 1945.

M-2 PARACHUTIST HELMET

The modified M-1 helmet adopted by paratroops differed from the standard M-1 helmet assembly in that the steel body's web chin strap was extended and incorporated a male snap fastener, which allowed the body to be secured to the liner during the jump. This modification prevented the separation of the liner from the helmet body. Modifications to the M-1 liner included the addition of female snap fasteners to receive the helmet body's web chin strap, inverted khaki, olive drab shade no. 3, cotton webbed A-straps with wire buckles, and a chamois lined leather molded chin cup. The inverted khaki, olive drab shade no. 3, cotton webbed A-straps were secured to the liner by removing four of the suspension washer and rivet assemblies, sandwiching the ends of the A-straps between the suspension washer and rivet assembly, and riveting the assembly together again. The buckles of the webbed strap allowed for adjustment of the chin cup.

Orders for the parachutist helmet were placed in January 1942, however, specifications for the new Parachutist Helmet were not written until June 1942. When the specification was finally written in June 1942, the designation given to the new Parachutist Helmet was the M-2. The Specification for the M-2 Parachutist Helmet called for the use of fixed D ring chin strap loops in place of the standard loops. These loops would allow the web chin strap to be worn more easily behind the wearer's head. Later, around the spring or summer of 1943, the Airborne Command stated that the standard chin strap loops were satisfactory and the fixed D ring chin strap loops were no longer warranted. In October 1943 production changed to a helmet with hinged chin strap loops. Between January 1942 and December 1944, approximately 148,000 helmet bodies

Interior view of the parachutist plastic helmet liner used with the M-1C, ca. January 1945-August 1945. The liner was manufactured by the Westinghouse Electric Company. (Courtesy of Keith R. Jamieson, M.D.)

were taken from existing McCord M-1 stocks and modified by McCord to parachutists configuration.

The modified parachutist helmet liners were also taken from existing stocks. The modified parachutist helmet liners retained the designation M-1. Initially, 43,000 Hawley fiber liners were modified between January 1942 and fall of 1942 by the McCord Radiator and Manufacturing Company. During the fall of 1942, 75,000 Inland Manufacturing plastic liners were also modified by the McCord Radiator and Manufacturing Company. From September 1943 until spring of 1944, Westinghouse supplied and modified the last major batch of the parachutist helmet liners. Approximately 30,000 helmet liners were modified by Westinghouse.

In the spring of 1944, the chamois lined leather molded chin cup was replaced by a khaki, olive drab shade no. 3, webbed chin strap cup for both production and probable replacement. The change was made due to the cheaper manufacturing cost of the web chin strap cup and approximately 40,000 web chin strap cups were produced during the spring of 1944.

M-1C PARACHUTIST HELMET

In January 1945, the M-1C Helmet was standardized by the U.S. Army. The M-1C was essentially the same as the earlier M-2 Parachutist Helmet and also incorporated all of the other changes found in the standard M-1 helmet. Between January 1945 and August 1945, the U.S. Army procured 392,000 M-1C Helmets.

Slight differences in the M-1C helmets included the use of olive drab shade no. 7 webbing in the manufacture of the liner A-straps, and the use of A-straps with cast buckles.

Both the M-2 and M-1C weighed 3 pounds, 2 ounces. The helmets weighed more than their standard counterparts due in part to the addition of the extra components.

PARACHUTIST HELMET TOTALS, 1942 - 1945

It is probable that an equal number of helmet bodies were produced or modified for every parachutist helmet liner. The total amount of parachutist helmets procured during World War II was approximately 540,000, with more than half being M-1Cs.

M-1 PARACHUTIST HELMET, 1955 - 1962

In 1955 production parachutist helmet liners were once again warranted. These production parachutist helmet liners were manufactured by taken existing production helmet liners and modifying them in the same method as was done during World War II. Female snap fasteners were added to receive the helmet body's web chin strap, and inverted olive drab shade no. 7, cotton webbed A-straps with wire buckles, and webbed chin

Close-up of the flexible chin strap loop and olive drab shade no. 7 webbed chin strap on the M-1 parachutist helmet body, ca. 1955-1960.

Exterior front left view of the M-1 parachutist helmet liner, ca. 1955-1962. This particular liner was manufactured in 1958.

Interior view of the M-1 parachutist helmet liner, ca. 1955-1962. This particular liner was manufactured in 1958. Note method of A-strap attachment.

Interior view of the M-1 parachutist helmet body, ca. 1966-1988. (Courtesy of Larry Sutherland)

Exterior view of the M-1 parachutist helmet body, ca. 1966-1988. (Courtesy of Larry Sutherland)

Exterior view of the M-1 parachutist helmet liner, ca. 1966-1968. This particular liner was manufactured in 1967. (Courtesy of Larry Sutherland)

strap cup, were added to the liner. The inverted A-straps were secured to the liner by removing four of the suspension A washer and rivet assemblies, sandwiching the ends of the A-straps between the suspension washer and rivet assembly, and riveting the assembly together again.

The parachutist helmet body was again in production between 1955 and 1962. Production of this item only required the addition of the body's extended web chin strap to existing production helmet bodies. This was relatively easy to accomplish as the helmet body's chin strap was now made removable by use of the chin strap clip.

M-1 PARACHUTIST HELMET, 1966 - 1968

In 1966 the last variation of the M-1 parachutist helmet liner was produced. Existing production nylon helmet liners were taken and modified to parachutist standards through the addition of female snap fasteners to receive the helmet body's web chin strap, and inverted olive drab shade no. 7, cotton webbed A-straps with wire buckles, and webbed chin strap cup. The A-straps were secured to the liner by use of four additional rivets located between the suspension A washers and the liner's rim. This type of installation did not require the disassembly of the liner's suspension.

The parachutist helmet body required the addition of the body's extended web chin strap to existing production helmet bodies.

Although a new chin strap was introduced in 1973 for the Ground Troops version of the M-1 helmet, the Parachutist version of the M-1 helmet continued to use the 1966 parachutist helmet body chin strap.

Exterior view of the M-1 parachutist helmet liner, ca. 1966-1968. This particular liner was manufactured in 1967. (Courtesy of Larry Sutherland)

View of the last variation M-1 parachutist helmet boby and liner, ca. 1966-1988. (Courtesy of Larry Sutherland)

Interior view of the M-1 parachutist helmet liner, ca. 1966-1968. This particular liner was manufactured in 1967. Note method of A-strap attachment. (Courtesy of Larry Sutherland)

U.S. Army parachutist wearing the last version M-1 parachutist helmet, ca. 1968.

5

THE M-3, M-4, M-4A2 AND M-5 AIRCREW HELMETS

On 9 August 1943, a project was initiated to provide a set of steel aircrew helmets for the United States Army Air Forces. Development of these aircrew helmets was the responsibility of the Army Air Forces at Wright Field and the Army Ordnance Department. All of these helmets had to allow for the use of the wearing of headphones, goggles and oxygen mask.

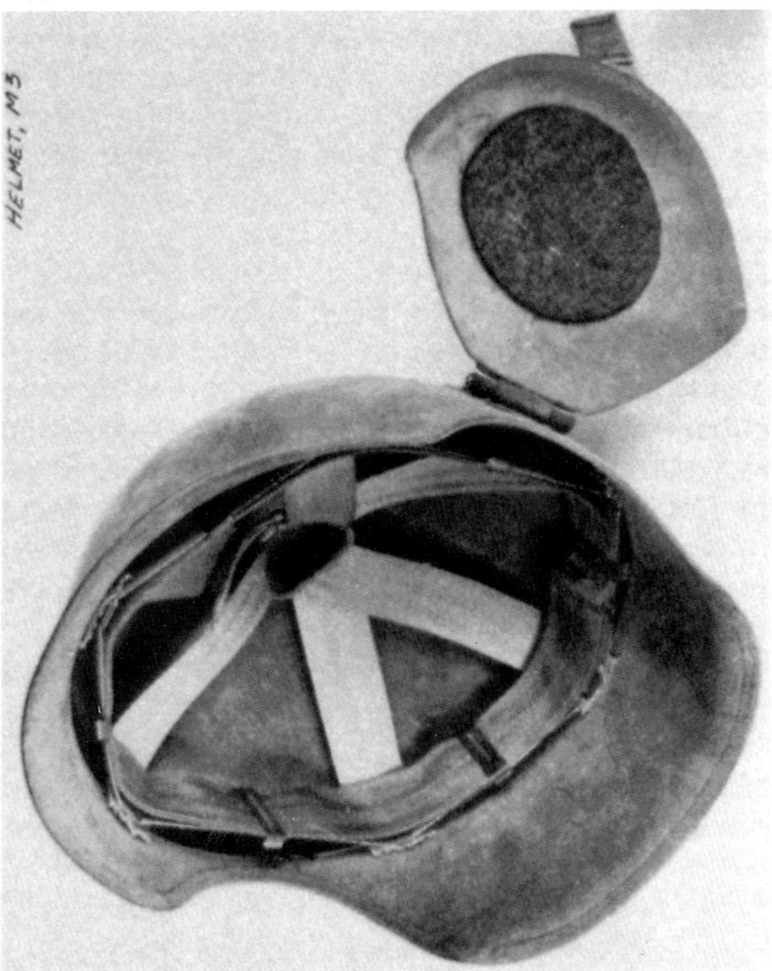

Interior view of the M-3 helmet, ca. December 1943-fall 1944. (Courtesy of Office of the Chief of Ordnance, U.S. Army)

M-3 HELMET

On 30 September 1943, a modification of the M-1 helmet, designated T-2, was judged satisfactory for most combat aircrew positions. As a result the T-2 was placed on limited production status and these helmets were subjected to further service tests. By December 1943 the T-2, was standardized as the Helmet, M-3.

The M-3 was a modification of the M-1 helmet. The differences being that no liner was provided, as the suspension was attached directly to the steel helmet body. A head band of cotton Herringbone twill was attached via spring clips to the suspension. The sides of the helmet body were cut away to provide clearance for the headphones. Hinged steel ear plates were spot welded to the helmet body to provided protection to the headphones.

The M-3 helmet shown integrated with other Air Force headgear accessories, February 1944. (Courtesy of U.S. Army Air Force)

Exterior right front view of the M-3 helmet, ca. December 1943-fall 1944.

Exterior front left view of the M-3 helmet, ca. December 1943-fall 1944. (Courtesy of Jon A. Maguire)

Interior view of the M-3 helmet, ca. December 1943-fall 1944. (Courtesy of Jon A. Maguire)

Close-up interior view of the hinged steel ear plate on the M-3 helmet, ca. December 1943-fall 1944. Note the rusted area where the round felt pad was glued in place.

Close-up view of the flexible chin strap loop and olive drab shade no. 3 chin strap on the M-3 helmet, ca. December 1943-fall 1944.

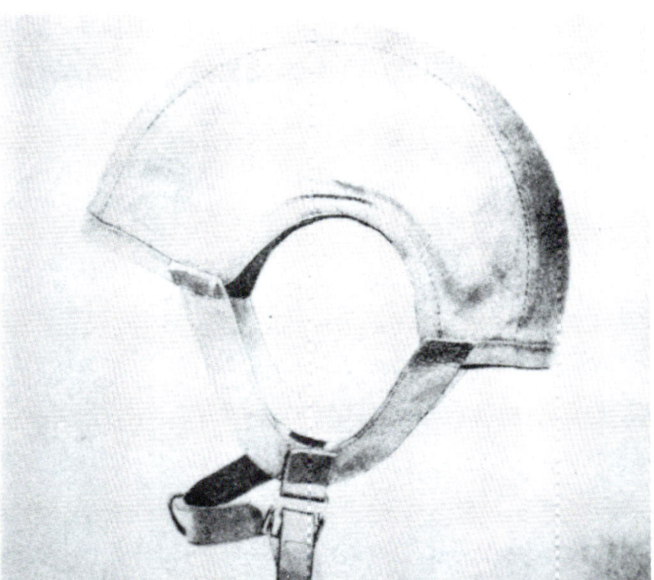

Exterior side view of the M-4 helmet, ca. December 1943-spring 1944. (Courtesy of Office of the Chief of Ordnance, U.S. Army)

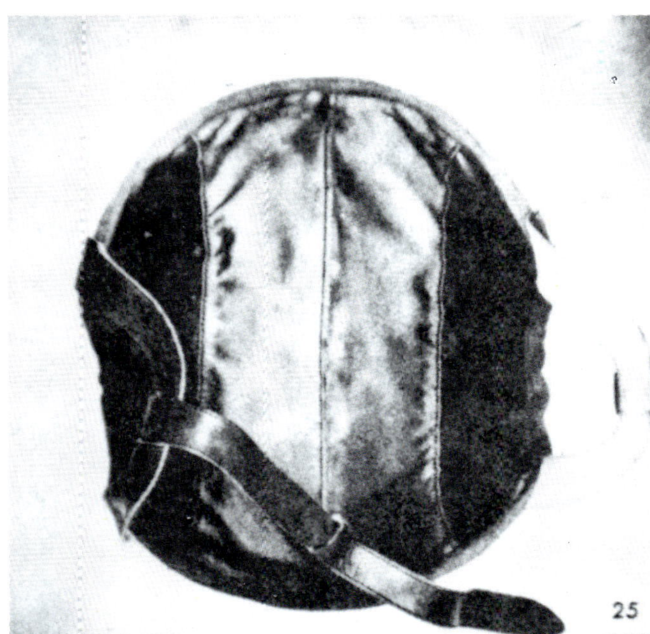

Interior view of the M-4 helmet, ca. December 1943-spring 1944. (Courtesy of Office of the Chief of Ordnance, U.S. Army)

The M-4 helmet shown integrated with other Air Force headgear accessories, February 1944. (Courtesy of U.S. Army Air Force)

Glued to the interior of the ear plates were round felt pads. A two piece cotton web, olive drab shade no. 3, chin strap was used to secure the helmet to the wearer's head. The chin strap was attached to the ear plates by use of flexible chin strap loops. From the fall of 1944 onwards the two piece cotton web chin strap was produced in olive drab shade no. 7. The M-3 helmet was painted an olive drab shade inside and out, with the exterior receiving a coat of flock to prevent bare skin from freezing to the helmet at altitude. The M-3 weighed 3 pounds, 3 ounces.

The McCord Radiator and Manufacturing Company of Detroit, Michigan, was selected by the U.S. Army to produce the M-3. McCord contracts indicate 194,683 M-3 helmets were delivered, while U.S. Army Ordnance procurement figures for the M-3 helmets during World War II, were as follows:

Year	Quantity
1943	2,613
1944	166,540
1945	44,390

Exterior front left view of the M-4 helmet, ca. December 1943-spring 1944. (Courtesy of Jon A. Maguire)

Exterior front right view of the M-4 helmet, ca. December 1943-spring 1944. (Courtesy of Jon A. Maguire)

Interior view of the M-4 helmet, ca. December 1943-spring 1944. (Courtesy of Jon A. Maguire)

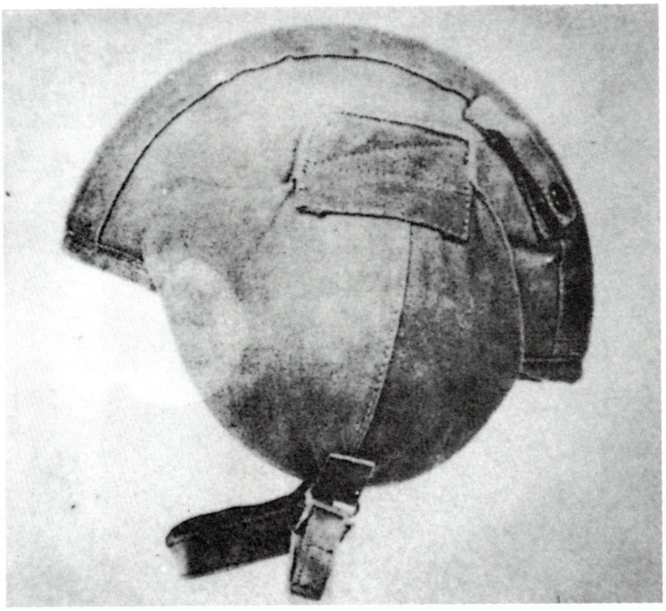

Exterior side view of the M-4A2 helmet, ca. spring 1944-August 1945. (Courtesy of Office of the Chief of Ordnance, U.S. Army)

Exterior left front view of the M-4A2 helmet, ca. spring 1944-August 1945.

Exterior right rear view of the M-4A2 helmet, ca. spring 1944-August 1945.

Interior view of the M-4A2 helmet, ca. spring 1944-August 1945.

M-4 HELMET

While the T-2 was judged satisfactory for most combat aircrew positions, it was not satisfactory for other cramped aircrew positions. Thus a second design, designated the T-3, was tested with the T-2 in the fall of 1943 at Wright Field. The T-3, was of a design that had been previously tested by 8th Air Force. This helmet could be used in those cramped aircrew positions where the M-3 was too large. The T-3 completed its service test with the T-2, and was standardized as the Helmet, M-4, on 2 December 1943.

The M-4 was smaller than the M-3, and thus provided less protection. The M-4 was basically a skull cap. The M-4 was made of Hadfield Manganese steel cut in strips. These strip were overlapped by placing them in separate fabric pockets which formed the cover of the M-4 helmets. The M-4 was covered in dark brown leather and secured to the wearer by use of an adjustable leather chin strap. The M-4 had no ear plates. The M-4 weighed 2 pounds, 1 ounce.

M-4A2 HELMET

The M-4A2 was an improvement over the M-4A1. The M-4A1 was basically the M-4 with increased protection in the form of ear plates. Additionally the M-4A1 was covered with cloth instead of leather. The ear plates were sewn on to the cloth covered helmet body.

Approximately 0.375" of length were added to increase protection with the M-4A2. A new method of securing the ear plates to the cloth body was incorporated. Additionally, ease of manufacture, improved clearance of goggles and greater ease in securing and removing the helmet to the wearer's head were also provided. The Industrial Canvas Products Corporation was the contractor for the M-4A2 helmet.

M-4 and M-4A2 HELMET TOTALS

The McCord Radiator and Manufacturing Company of Detroit, Michigan, was selected by the U.S. Army as a sub-contractor to produce Hadfield Manganese steel strip sets for the M-4 and

Close-up of the identification tag found on the M-4A2 helmet, ca. spring 1944-August 1945. (Courtesy of Jon A. Maguire)

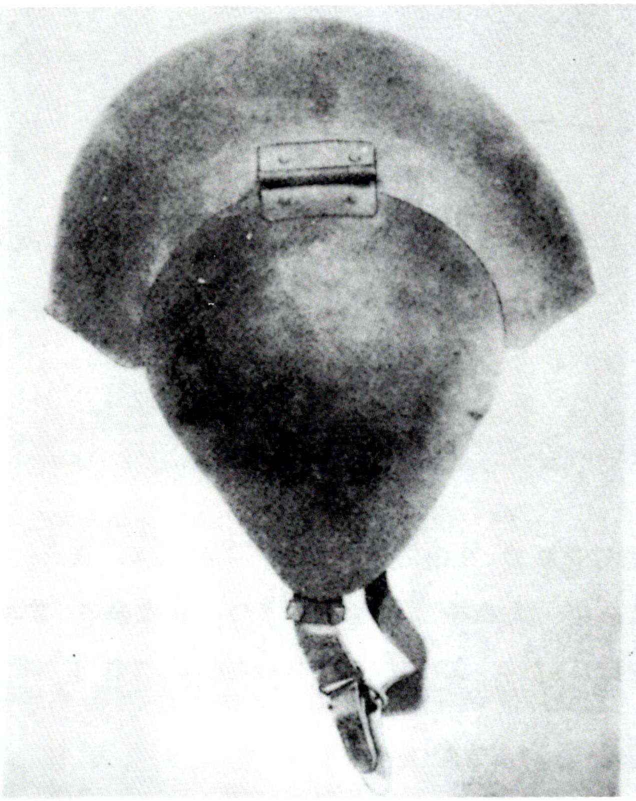

Exterior side view of the M-5 helmet, ca. January 1945-August 1945. (Courtesy of Office of the Chief of Ordnance, U.S. Army)

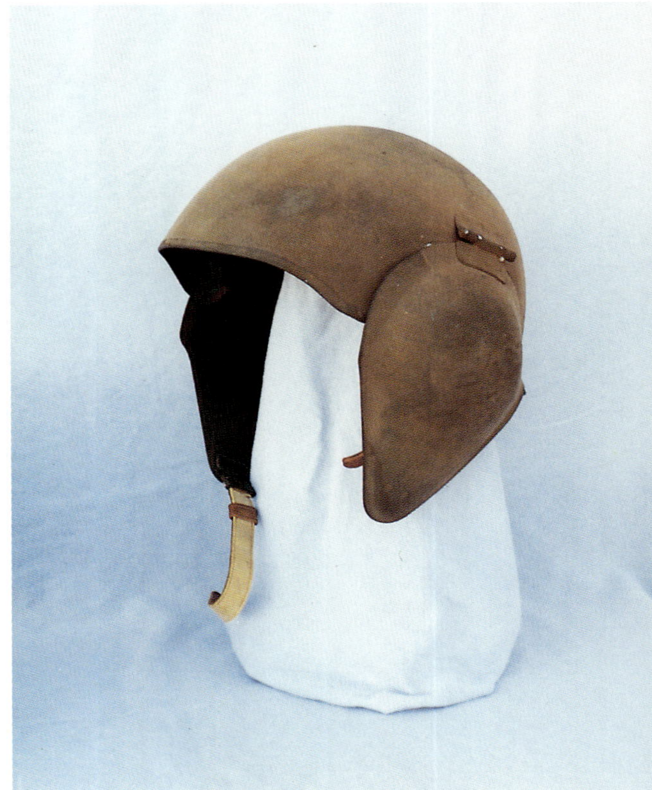

Exterior left front view of the M-5 helmet, ca. January 1945-August 1945.

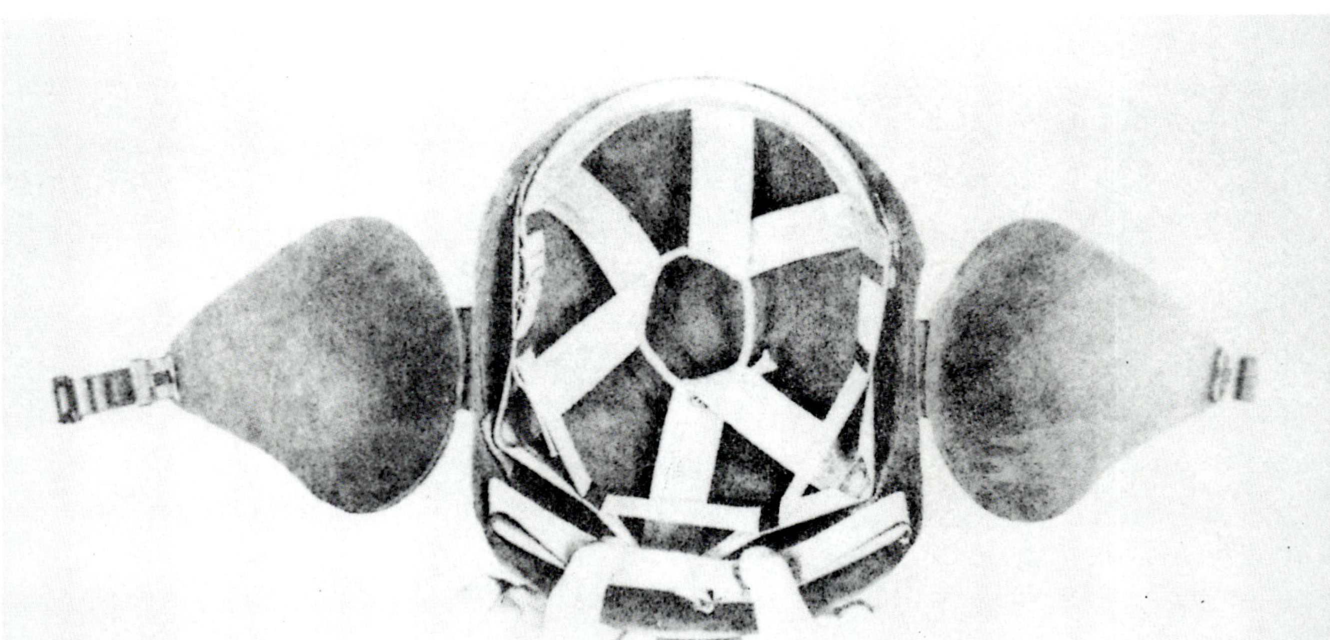

Interior view of the M-5 helmet, ca. January 1945-August 1945. Note adjustable suspension. (Courtesy of Office of the Chief of Ordnance, U.S. Army)

Close-up view of the flexible chin strap loop, chin strap fastener, and leather chin strap on the M-5 helmet, ca. January 1945-August 1945.

Interior view of the M-5 helmet, ca. January 1945-August 1945. (Courtesy of Jon A. Maguire)

A USAAF B-29 pilot wearing an M-5 helmet, June 1945. (Courtesy of U.S. Army Air Force)

M-4A2. McCord contracts indicated that 55,000 steel strip sets were delivered, while U.S. Army Ordnance procurement figures for the M-4 and M-4A2 helmets during World War II, were as follows:

Year	Quantity
1943	1,353
1944	54,542
1945	30,241

M-5 HELMET

In November 1943, development of an all-purpose Air Force steel helmet was initiated. From this early work came a helmet that was designated as the T-6E4. The T-6E4 helmet fitted closely to the wearer's head, similar to a skull cap. This bowl was manufactured from a single piece of steel and contained no suspension, as the distance between the wearer's head and the bowl ranged from 0.03125" to 0.25". Attached to the bowl were two hinged ear flaps. The ear flaps of the T-6E4 provided greater protection than those of the T-3. From this promising design came the T-8 helmet.

The T-8 helmet was based on the T-6E4, but included many enhancements. The helmet had a bowl with two attached check plates. The bowl and check plates followed the contour of the Army Air Force's leather flying helmets, such as the A-11. The check plates were embossed for the headphones of the leather helmet and also to provided clearance for the goggle straps when the goggles were worn on the brow of the steel helmet. A head suspension was provided in the T-8. The suspension was adjustable to fit various head sizes. An adjustable cotton webbing strap was attached at the back of the helmet. This strap was adjusted to ensure that the forehead of the wearer was held constantly against the front of the steel helmet, so as not to interfere with vision. The suspension was similar to the M-3, with all of the webbing being manufactured in cotton, olive drab shade no. 3, twill. Additionally, a leather brow pad was provided to increase comfort to the wearer's forehead. The helmet was secured by use of a removable leather chin strap. The chin strap was similar to those used for the A-11 helmet, and was made removable through use of a chin strap fastener.

The T-8 was evaluated and tested by the Aero-Medical Laboratory at Wright Field, and because of its significant progress, development of all other Air Force steel helmets was stopped. The helmet was then studied and evaluated by the Air Ordnance Officer, Army Air Forces Board at Orlando, Florida. A report from the Board recommended that the helmet be standardized so long as the rear lower edge was rolled to minimize neck injuries in a crash landings. In January 1945, the T-8 helmet was standardized as the Helmet, M-5. The M-5 helmet was to be used in all combat aircraft crew positions except the Martin upper turret gunner of the A-20 and the ring-sight gunner of the B-29. The M-4A2 helmet continued to be standard for these two unique gunner positions. The M-5 helmet weighed 2 pounds, 12 ounces.

The McCord Radiator and Manufacturing Company of Detroit, Michigan, was selected by the U.S. Army to produce the M-5. The U.S. Army Ordnance Department procured 93,495 M-5 helmets between January 1945 and August 1945.

6

THE STEEL HELMET Mk. II

In 1940, the U.S. Navy's Bureau of Ordnance completed development of a telephone talker's helmet for certain ex posed shipboard personnel. This helmet was designated as the Steel Helmet Mk. II. This helmet was also referred to as the "Talker's" helmet or "Special Navy Type" helmet.

The McCord Radiator and Manufacturing Company of Detroit, Michigan, was selected by the U.S. Navy to produce the the Steel Helmet Mk. II. The helmet was manufactured from a single piece of Hadfield Manganese, and made in only one size. The Steel Helmet Mk. II provided more protection to the forehead , sides and back of the head than did the U.S. Army's M-1917A1 helmet then in use. The Steel Helmet Mk. II was of one piece construction with an inner support of soft sponge, latex or neoprene, cemented to the inside of the steel shell. These soft sponge supports were known as Foamex liners and manufactured by the Firestone Tire & Rubber Company. The Foamex pad itself was covered with a grey/blue cloth. The Foamex liners improved the acoustic properties of the helmet. A leather chin strap with chin cup was secured to the shell with wire links. The leather chin cup was lined with retanned horsehide which remained comfortable in severe weather conditions. Originally the Steel Helmet Mk. II was painted with a lusterless olive drab with ground cork as the exterior aggregate. In 1942, production changed and the helmets were painted deck blue with ground cork as the exterior aggregate for camouflage purposes. Between 1940 and 1945 the McCord Radiator and Manufacturing Company produced 400,000 Mk. II helmets for the U.S. Navy.

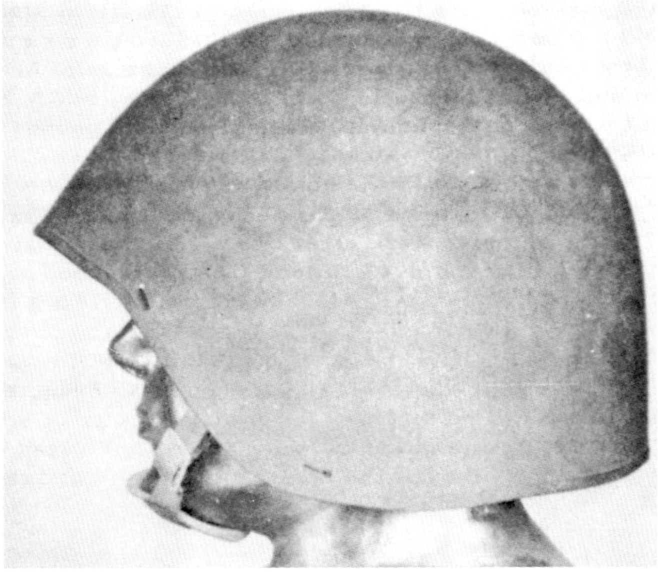

Exterior view of the Steel Helmet Mk. II. (Courtesy of Naval Medical Research and Development Command)

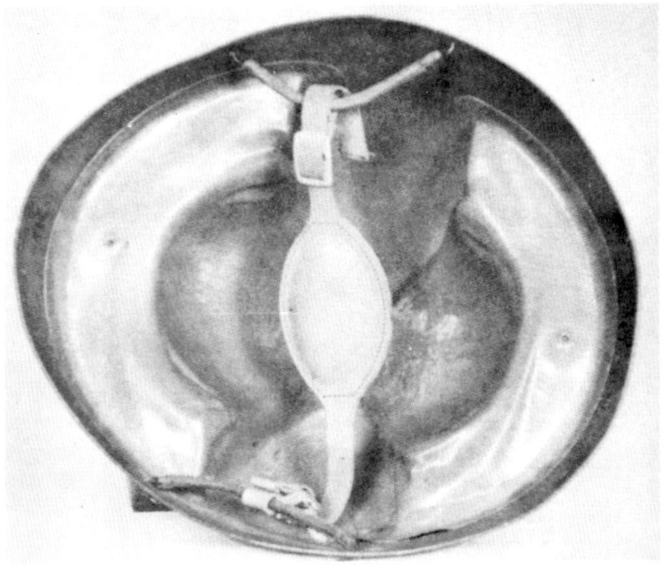

Interior view of the Steel Helmet Mk. II. (Courtesy of Naval Medical Research and Development Command)

Exterior view of the Steel Helmet Mk. II, ca. 1942-1945.

Interior view of the Steel Helmet Mk. II, ca. 1942-1945.

Front view of the Mk. II helmet chin strap, ca 1940-1945.

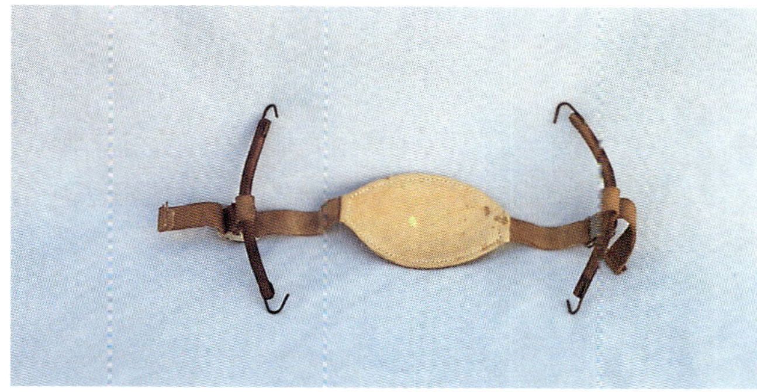

Rear view of the Mk. II helmet chin strap, ca 1940-1945.

Close-up of the Mk. II helmet exterior finish, ca. 1942-1945. This view shows the ground cork aggregate applied to the deck blue exterior paint.

Mk. II helmet interior visor stamp marking, ca. 1942-1945.

7

THE PASGT HELMET

In 1972 development of a new helmet to replace the M-1 helmet began. Develop of the new helmet was undertaken by the U.S. Army Natick Research, Development, and Engineering Command. The U.S. Army evaluated numerous ballistic materials for the construction of the new helmet, including fiberglass and aramid fiber. Both of these materials provided effective protection against fragmenting munitions. The Army Ballistic Research Laboratories and the Naval Research Laboratory finally selected the KEVLAR® brand fiber, the DuPont tradename for their aramid fiber material. By 1978 the U.S. Army had completed tests on the new helmet at the Human Engineering Laboratory as well as other Army Labs, and had standardized it along with a vest as the Personnel Armor System Ground Troops, PASGT. These tests ensured that the helmet was compatible with the common positions assumed by the solider in combat. The helmet was referred to as the PASGT Combat Helmet.

The PASGT combat helmet was constructed of 17 layers of aramid fiber material, and improved coverage of the head by 11 percent over the M-1 helmet. The layers of aramid fiber where built up and resin laminated in a mold. The PASGT helmet originally came in four sizes, that fit 98 percent of all U.S. Army personnel. These sizes being, extra small, small, medium, and large. The medium size fits approximately 50 percent of all soldiers. The interior of the helmet contained a cradle suspension system, much like the early plastic M-1 liner's, with a hook and pile pull-tab drawstring replacing the old shoestring. The drawstring allowed for the height adjustment. The suspension was secured to the helmet via a screw and A nut assembly. The helmet contained a head band, much like the M-1, with a adjustment buckle. The head band had leather sewn to one side of the band along a single edge. All webbed items of the suspension system were made of nylon webbing. The PASGT helmet contained no neck or nape strap. The helmet

Exterior right front view of the PASGT helmet, ca. 1978-1996.

Exterior left rear view of the PASGT helmet, ca. 1978-1996.

Exterior rear view of the PASGT helmet, ca. 1978-1996.

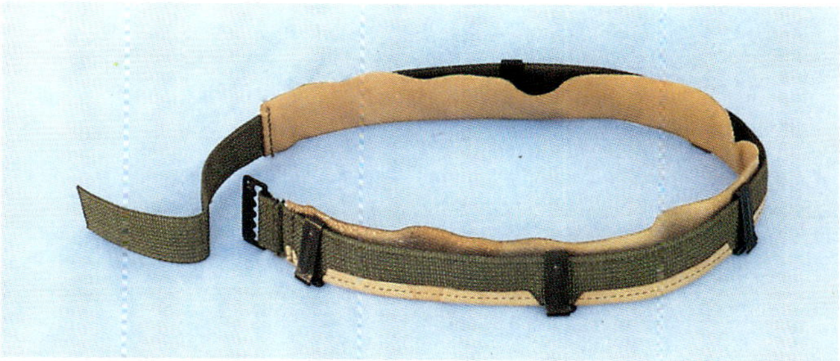

PASGT head band, ca. 1978-1996.

PASGT head band spring clip, ca. 1978-1996.

had a chin strap with a two point open web chin cup, two adjustment buckles and a single snap fastener on the left side. Initially the chin strap was made completely of cotton, later it was made of cotton and nylon. All webbing, whether cotton or nylon was of olive drab color. The helmet weighed a little more than 3 pounds. The PASGT helmets were manufactured and painted on the exterior an olive drab color with sand aggregate.

In February 1989, an additional sized helmet was added to production and was extra large.

Between 1978 and 1988, approximately 3,000,000 PASGT helmets were produced. The PASGT combat helmet was manu-

Interior left front view of the PASGT helmet, ca. 1978-1996.

Interior right rear view of the PASGT helmet, ca. 1978-1996.

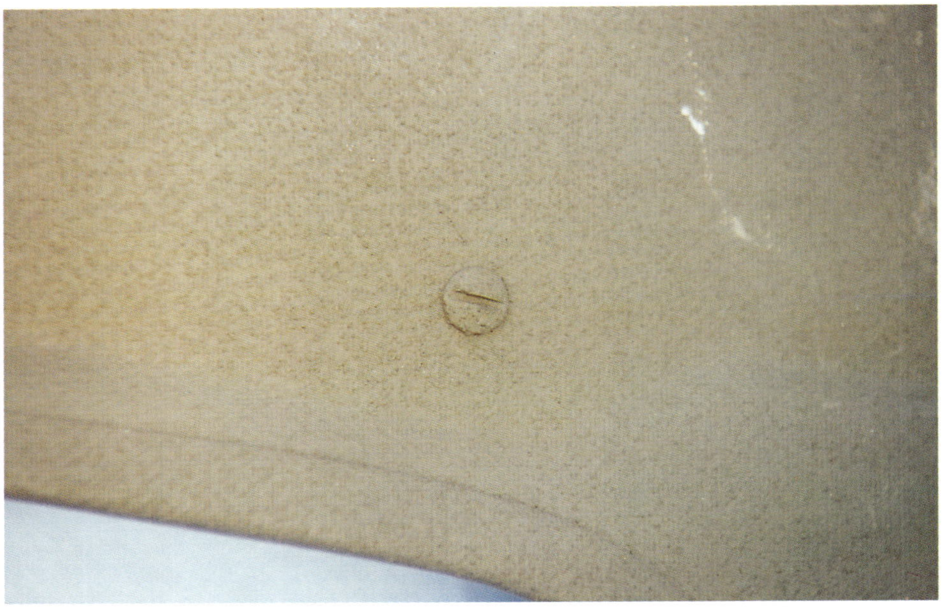

(Above) Close-up of screw on the PASGT helmet, ca. 1978-1996.

PASGT suspension "A" nut, ca. 1978-1996.

Close-up of stamp marking, mold-in marking, on the PASGT helmet, ca. 1978-1996.

Close-up of stamp marking on the PASGT helmet, ca. 1978-1996.

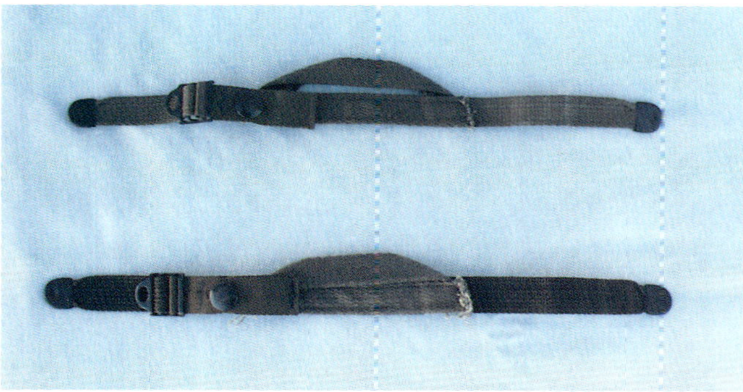

Exterior view of the PASGT chin strap. Top, early variation made of cotton, bottom, late variation made of nylon.

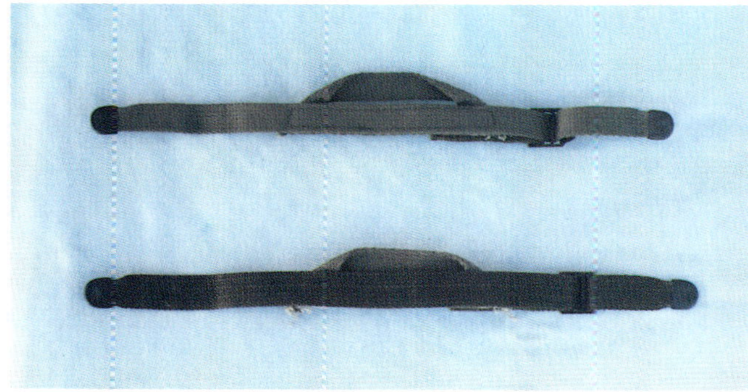

Interior view of the PASGT chin strap. Top, early variation made of cotton, bottom, late variation made of nylon.

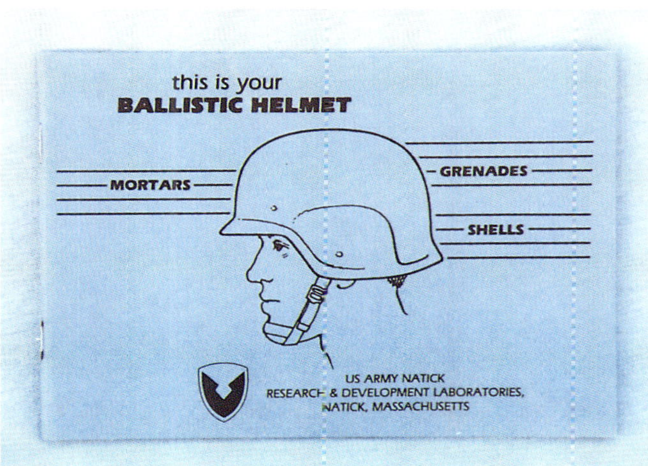

PASGT helmet booklet, ca. 1978-1996.

Interior view of the PASGT helmet complete, ca. 1978-1996.

PASGT camouflage helmet cover, "Woodland" pattern, ca. 1978-1996. (Courtesy of Anthony C. Wilsbacher)

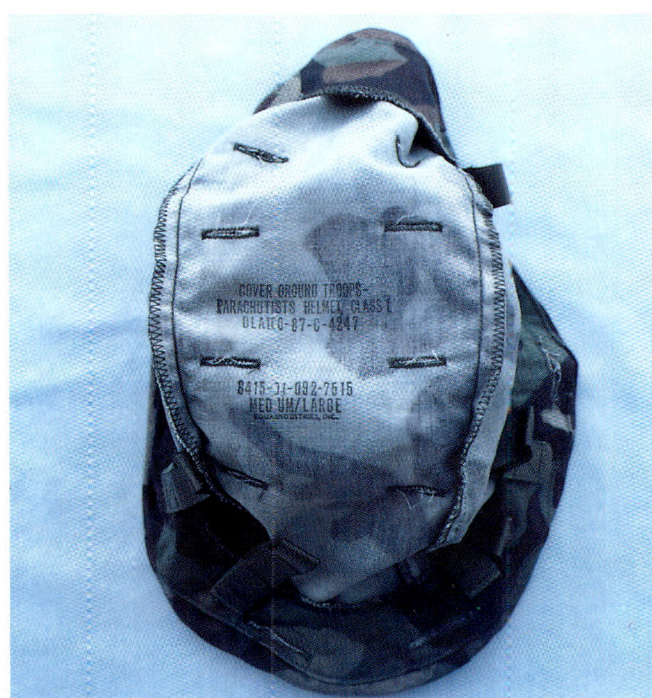

Interior view of the PASGT helmet cover, "Woodland" pattern, ca. 1978-1996. Note the stamp marking. (Courtesy of Anthony C. Wilsbacher)

Front left view of PASGT helmet with camouflage helmet cover, "Woodland" pattern, ca. 1978-1996.

Rear right view of PASGT helmet complete with "Woodland" pattern helmet cover, ca. 1978-1996.

Front view of PASGT helmet complete with "Woodland" pattern helmet cover, ca. 1978-1996.

Front left view of PASGT helmet complete with "Woodland" pattern helmet cover, ca. 1978-1996.

PASGT camouflage helmet cover, "Snow" pattern, ca. 1978-1996.

Front right view of PASGT helmet with camouflage helmet cover, "Snow" pattern, ca. 1978-1996.

PASGT camouflage helmet cover, "Desert Daytime" pattern, ca. 1978-1996. (Courtesy of Anthony C. Wilsbacher)

Side view of PASGT helmet with camouflage helmet cover, "Desert Daytime" pattern, ca. 1978-1996.

Front right view of PASGT helmet complete with "Desert Daytime" pattern helmet cover, ca. 1978-1996.

Rear left view of PASGT helmet complete with "Desert Daytime" pattern helmet cover, ca. 1978-1996.

Interior view of PASGT helmet and helmet cover. This view shows the method of attachment of the helmet cover to the helmet. Note the hook and pile attachment straps.

U.S. Army soldier wearing the PASGT helmet complete with the "Desert Daytime" cover. (Courtesy of U.S. Army)

1LT Anthony Wilsbacher on the berms on the border of Saudia Arabia and Iraq, February 1991. Just WNW of border town of Nisab. Note PASGT helmet and "Desert Daytime" helmet cover. (Courtesy of Anthony C. Wilsbacher)

Front view of the PASGT parachutist pad and retention strap. (Courtesy of Chris Armold)

Rear view of the PASGT parachutist pad and retention strap. (Courtesy of Chris Armold)

Interior view of the PASGT parachutist helmet. Black foam crown pad non-standard.

factured by the Gentex Corporation, Carbondale, Pennsylvania; Specialty Plastic Products, Scranton, Pennsylvania; Sioux Manufacturing, Fort Totten, North Dakota; and Unicor Corporation Division of Federal Prison Industries, Bastrop, Texas.

The initial production contracts for the PASGT helmet were placed with the Gentex Coporation in the late 1970's. Sioux Manufacturing was given successive contracts to produce the PASGT helmet from 1980 until 1991, and produced a total of 900,000 PASGT Helmets. In 1982, Unicor Corporation was awarded a production contract, and recieved additional contracts, on and off, through 1996. Production of PASGT helmets at Unicor started in 1983, and between 1983 and 1996

Exterior left side view of the PASGT parachutist helmet.

Exterior rear view of the PASGT parachutist helmet.

Unicor produced 2,500,000 PASGT Helmets. Specialty Plastic Products was also awarded contracts, and produced the helmet from the 1985 through 1996. In that time, they produced a total of 400,000 PASGT Helmets for U.S. Forces.

The U.S. Army required mold-in crown markings for all PASGT as well as a printed identification marking in ink at the rear of the inside surface of the helmet. The printed indentification included the description of the item, the National Stock Number and the manufacturer's name. The mold-in crown markings were: the word Gentex for Gentex Corporation; the letters SPP for Specialty Plastic Products; a horse and lighting bolt within a circle for Sioux Manufacturing; and four arranged blocks for Unicor Corporation.

THE PASGT PARACHUTIST HELMET

For parachutists the helmet had two additional pieces that could be fastened to the suspension system for support during parachuting operations, the Parachutist's Helmet Retention Strap and the Parachutist's Helmet Pad. The retention strap consisted of olive drab shade no. 7 nylon webbing with an attachment brass eyelet in the center and a hook and pile tape on each end of the strap. The pad was a trapezoidal shaped piece of vinyl nitrile foam which was attached to the rear of the helmet.

THE PASGT HELMET COVERS

There were three covers adopted for use with the PASGT combat helmet. These included the Woodland, Snow and Desert Daytime Camouflage Patterns. The covers were to make the PASGT helmet less conspicuous in the three different environments. The covers were made of a cotton/nylon twill fabric and included reinforced holes for the insertion of vegetation or other camouflage materials. The Woodland pattern contained a four color woodland camouflage pattern applied to one side of the material. The Snow pattern was a bleached white. The Desert Daytime pattern was a six color desert camouflage pattern applied to one side of the material. The covers came in two sizes, X-Small/Small and Medium/Large. There were no covers available for the X-Large PASGT helmet. The cover were attatched to the helmet suspension by means of hook and pile fasteners on the end of nylon strips.

Among the firms who produced the cover were Equa Industries; Medical Supplies Manufacturing; Royal Maid, Topelo, Mississippi; and Volunteer Blind Industries.

BIBLIOGRAPHY

Chapter 1. The M-1917 Helmet

Crowell, Benedict, "America's Munitions", 1917 -1918, Report of, United States Army War Office, Washington D.C. 1919

Lewis, Frederick J., Commander, MSC, USN, et al, "Military Helmet Design", Naval Medical Field Research Laboratory, Camp Lejeune, North Carolina, June 1958

Chapter 2. The M-1917A1 Helmet

Lewis, Frederick J., Commander, MSC, USN, et al, "Military Helmet Design", Naval Medical Field Research Laboratory, Camp Lejeune, North Carolina, June 1958

Massen, Marion, "History of the Helmet Liner", C.Q.M.D. Historical Studies: Report No. 5, Chicago Quartermaster Depot, ASF, Chicago, Illinois, April 1944

McCord Write-up, "Historical Record of McCord Corporation Contribution to World War II Ordnance Material", Detroit, Michigan, August 1945

Chapter 3. The M-1 Helmet

Lewis, Frederick J., Commander, MSC, USN, et al, "Military Helmet Design", Naval Medical Field Research Laboratory, Camp Lejeune, North Carolina, June 1958

Lief, Alfred, "The Firestone Story", Firestone Tire and Rubber Company, Akron, Ohio, 1951

Massen, Marion, "History of the Helmet Liner", C.Q.M.D. Historical Studies: Report No. 5, Chicago Quartermaster Depot, ASF, Chicago, Illinois, April 1944

McCord Write-up, "Historical Record of McCord Corporation Contribution to World War II Ordnance Material", Detroit, Michigan, August 1945

"New Combat Helmets", Steel, Volume 111, Cleveland, Ohio, October 5, 1942

"Quartermaster Supply Catalog", Army Service Forces Catalog QM sec.1, OQMG Circular No. 4-Revised August 1943, Quartermaster General-ASF, Washington D.C., August 1943

"Record of Army Ordnance, Research and Development", Volume 2, Office of the Chief of Ordnance, Research and Development Service, Washington D.C., January 1946

Risch, Erna, "The Quartermaster Corps: Organization, Supply and Services", Volume 1, The United States Army in World War II, Technical Services, Office of the Chief of Military History, Washington D.C., 1953

Studler, Rene R., Colonel, USA, "The New Combat Helmet", Army Ordnance, Volume 22, Washington D.C., May - June, 1942

Thomson, Harry C., and, Mayo, Lida, "The Ordnance Department: Procurement and Supply", The United States Army in World War II, Technical Services, Office of the Chief of Military History, Washington D.C., 1960

Wells, Daniel L., "The Story of the New American Helmet", McCord Radiator and Manufacturing Company, Detroit, Michigan, ca.1946

Westinghouse Write-up, Bryant Electric, "Plastic Helmet Liners", Bridgeport, Connecticut, ca.1946

Westinghouse Write-up, Micarta Division, "Helmet Liners", Trafford, Pennsylvania, ca.1946

Whiting, "Statistics, United States Army in World War II", Office of the Chief of Military History, Special Staff, United States Army, Washington D.C., ca.1946

SPECIFICATIONS

Helmet Body

Helmet, Steel, M-1

O.D. AXS 1138	
O.D. AXS 1138 Rev	
O.D. AXS 1138 Rev	
MIL-H-10990	March 1951
MIL-H-10990	August 1958 CANC
MIL-H-1988	August 1958
MIL-H-1988A	
MIL-H-1988B	
MIL-H-1988C	
MIL-H-1988D	
MIL-H-1988E	April 1968
MIL-H-1988F	1975
MIL-H-1988G	March 1988

Ground Troops Chin Strap

Chin Strap, Ground Troop (Nylon) for Helmet, M-1

MIL-S-43841	February 1973
MIL-S-43841A	June 1976
MIL-S-43841A	September 1977

Helmet Liner

Lining, Helmet, Assembly for Helmet, Steel, M-1

O.D. AXS 644	October 1941

Liner, Helmet, M-1

OQMG No. 42	February 1942
C.Q.D. No. 65	June 1942
C.Q.D. No. 65a	July 1942
C.Q.D. No. 65b	August 1943

Liner, Helmet, M-1 (Impreganted Fabric, Cotton)

MIL-L-1910	January 1950
MIL-L-1910A	
MIL-L-1910B	September 1963
MIL-L-1910C	April 1966
MIL-L-1910D	April 1969
MIL-L-1910	February 1975 CANC

Liner, Helmet, M-1 (Nylon)

MIL-L-41800	
MIL-L-41800A	May 1963
MIL-L-41800B	July 1965
MIL-L-41800C	April 1966
MIL-L-41800D	March 1969
MIL-L-41800E	June 1974
MIL-L-41800F	April 1976
MIL-L-41800G	December 1985

Head Band

Head-Band, Assembly for Liner, Helmet, M-1

C.Q.D. No. 63	June 1942
C.Q.D. No. 63	June 1942
C.Q.D. No. 63a	July 1942
C.Q.D. No. 63b	October 1943

Band, Liner, Helmet, M-1, Head, New Type
C.Q.D. No. 63c — February 1944
C.Q.D. No. 63d — August 1944

Neck Band
Neck-Band, for Liner, Helmet, M-1
C.Q.D. No. 64 — June 1942
C.Q.D. No. 64a — July 1942

Neck-Band, for Liner, Helmet, M-1, Adjustable
C.Q.D. No. 64b — August 1944
U.S.A. No. 6-344 — March 1945

Head and Neck Band
Bands: Head and Neck, for Liner, Helmet, M-1
MIL-B-1953 — January 1950
MIL-B-1953A

Head Band and Neck Band, for Liner, Helmet, M-1
MIL-H-41802
MIL-H-41802A — September 1963
MIL-H-41802B — October 1966

Suspension
Suspension, Liner M-1 (Detachable)
MIL-S-43620 — January 1969
MIL-S-43620A — March 1976
MIL-S-43620B — August 1985

Accessories
Band, Helmet, Camouflage
P.Q.D. No. 255 — September 1942
P.Q.D. No. 255-a — October 1942
P.Q.D. No. 255-b — November 1942
P.Q.D. No. 255-c — April 1944
MIL-B-1851
MIL-B-1851A
MIL-B-1851B — August 1963
MIL-B-1851C — May 1967
MIL-B-1851D — July 1971
MIL-B-1851E — December 1976
MIL-B-1851F — May 1981

Cover, Helmet, Camouflaged
Marine Corps Spec. — September 1942
MIL-C-17502
MIL-C-17502A
MIL-C-17502B — June 1966
MIL-C-17502C — December 1971

Chapter 4. The M-2, M-1C and M-1 Parachutist Helmets

Lewis, Frederick J., Commander, MSC, USN, et al, "Military Helmet Design", Naval Medical Field Research Laboratory, Camp Lejeune, North Carolina, June 1958

"Record of Army Ordnance, Research and Development", Volume 2, Office of the Chief of Ordnance, Research and Development Service, Washington D.C., January 1946

SPECIFICATIONS

Parachutist Helmet Body
Helmet, Steel, (Parachutist), M-2
O.D. — June 1942
O.D. — November 1944

Helmet, Steel, (Parachutist), M-1C
O.D. — December 1944
O.D. Rev 1 — January 1945
O.D. Rev 2 — June 1945

Parachutist Helmet Liner
Liner, Helmet, (Parachutist), M-1
C.Q.D. No. 66 — July 1942

Parachutist Chin Strap
Chin Strap, Parachutist for Helmet, M-1
MIL-S-43912 — May 1975

Chapter 5. The M-3, M-4, M-4A2 and M-5 Aircrew Helmets

Lewis, Frederick J., Commander, MSC, USN, et al, "Military Helmet Design", Naval Medical Field Research Laboratory, Camp Lejeune, North Carolina, June 1958

McCord Write-up, "Historical Record of McCord Corporation Contribution to World War II Ordnance Material", Detroit, Michigan, August 1945

"Record of Army Ordnance, Research and Development", Volume 2, Office of the Chief of Ordnance, Research and Development Service, Washington D.C., January 1946

Chapter 6. The Steel Helmet Mk. II

Lewis, Frederick J., Commander, MSC, USN, et al, "Military Helmet Design", Naval Medical Field Research Laboratory, Camp Lejeune, North Carolina, June 1958

Lief, Alfred, "The Firestone Story", Firestone Tire and Rubber Company, Akron, Ohio, 1951

McCord Write-up, "Historical Record of McCord Corporation Contribution to World War II Ordnance Material", Detroit, Michigan, August 1945

Chapter 7. The PASGT Helmet

"Army Science Conference Paper Reports on Infantry Helmet Development", Army Reserch and Development News Magazine, March-April 1976

Lindblad, Jeff, "KEVLAR® in Vests, Helmets Saves Lives of Today's Soldiers", APG News, March 9, 1988

"New Infantry Helmet Designed for Options in 3 Sizes", Army Reserch and Development News Magazine, September-October 1976

Specialty Plastic Products Write-up, "Specialty Plastic Products, Inc. and its Technology", Dunmore, Pennsylvania, 1994

SPECIFICATIONS

Helmet Body
Helmet, Ground Troops - Parachutist's
MIL-H-44099 August 1978

Strap
Strap, Chin, Ground Troops - Parachutist's Helmet
MIL-S-44091 May 1978

Strap, Retention, Parachutist's Helmet
MIL-S-44022
MIL-S-44022A

Suspension
Suspension System, Helmet, Ground Troops - Parachutist's
MIL-S-44097 September 1977

Pad
Pad, Parachutist's Helmet
MIL-S-44081

Accessories
Cover, Ground Troops - Parachutist's Camouflage
MIL-C-44107
MIL-C-44107A

Band, Helmet, Camouflage
MIL-B-1851F May 1981

Appendix A:
Selected Military Specification Drawings

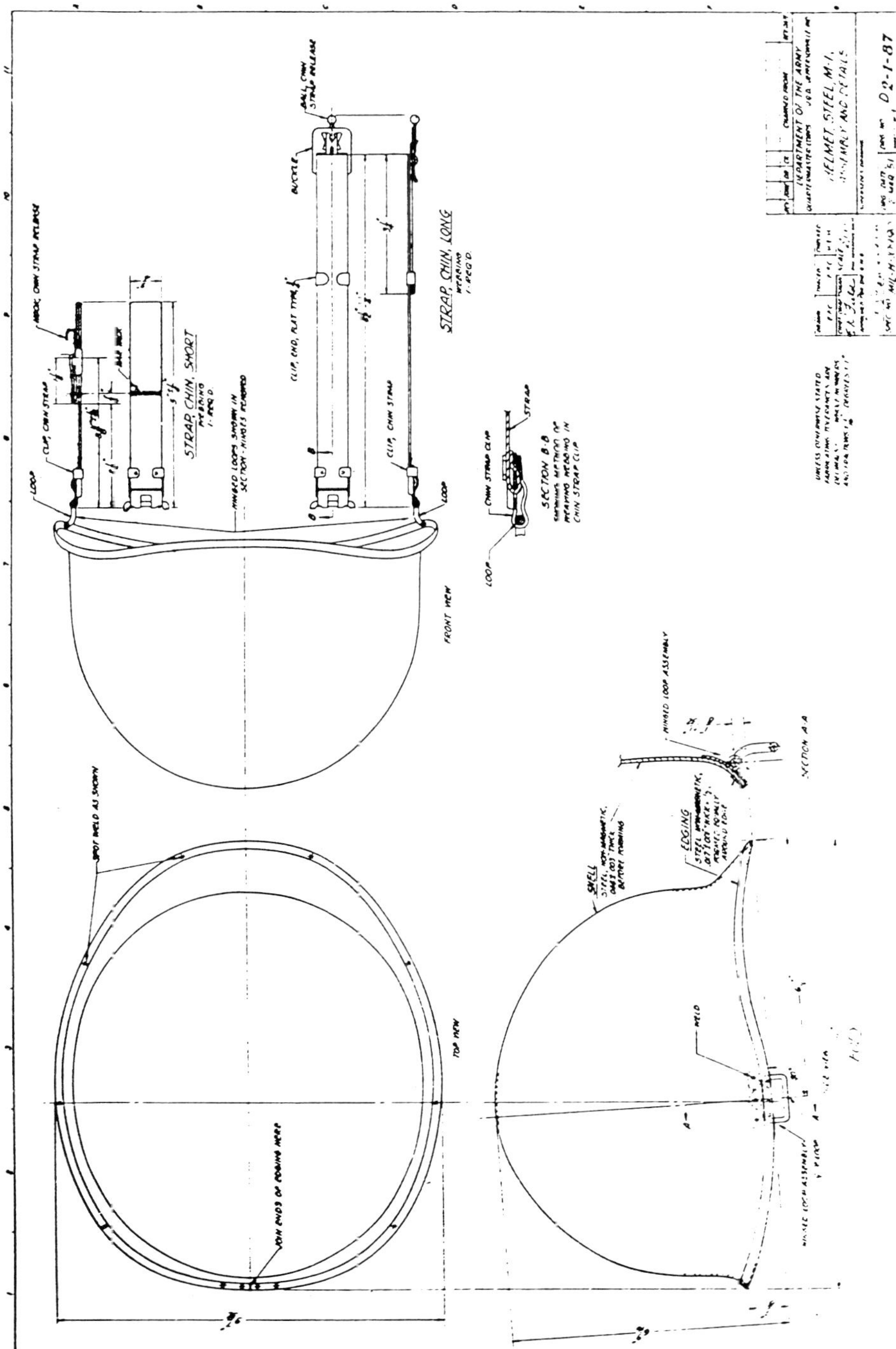

10.

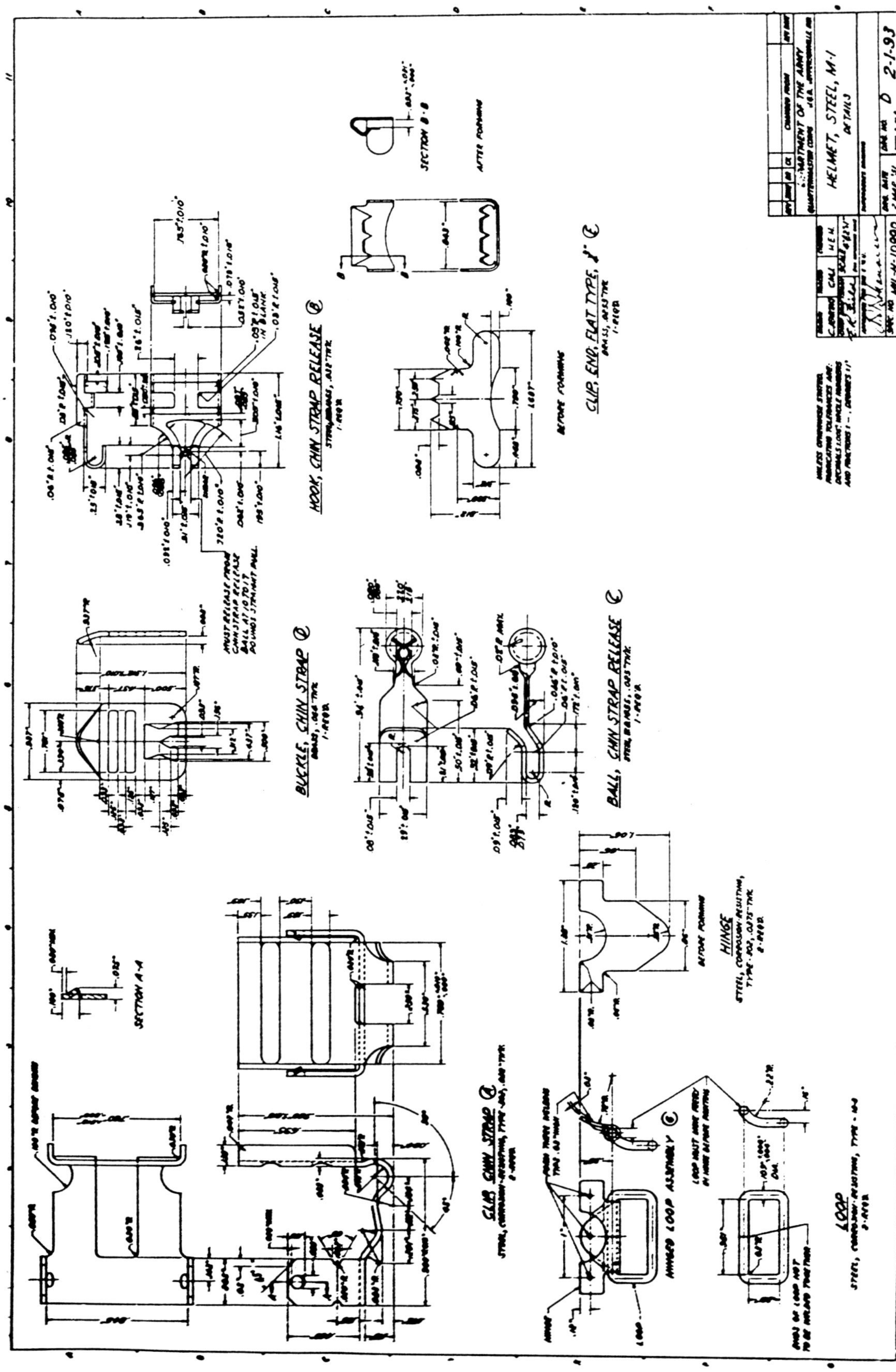
HELMET, STEEL, M-1
DETAILS
DEPARTMENT OF THE ARMY
D 2-1-93
HOOK, CHIN STRAP RELEASE
CLIP, END, FLAT TYPE
BUCKLE, CHIN STRAP
BALL, CHIN STRAP RELEASE
CLIP, CHIN STRAP
HINGE
HINGED LOOP ASSEMBLY
LOOP
SECTION A-A
SECTION B-B
BEFORE FORMING
AFTER FORMING

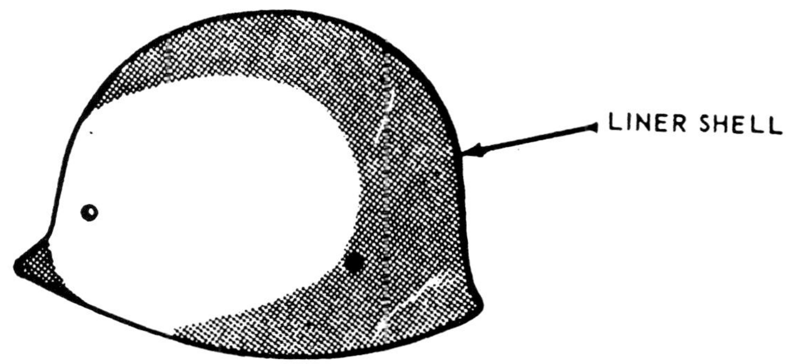

A. EXTERIOR

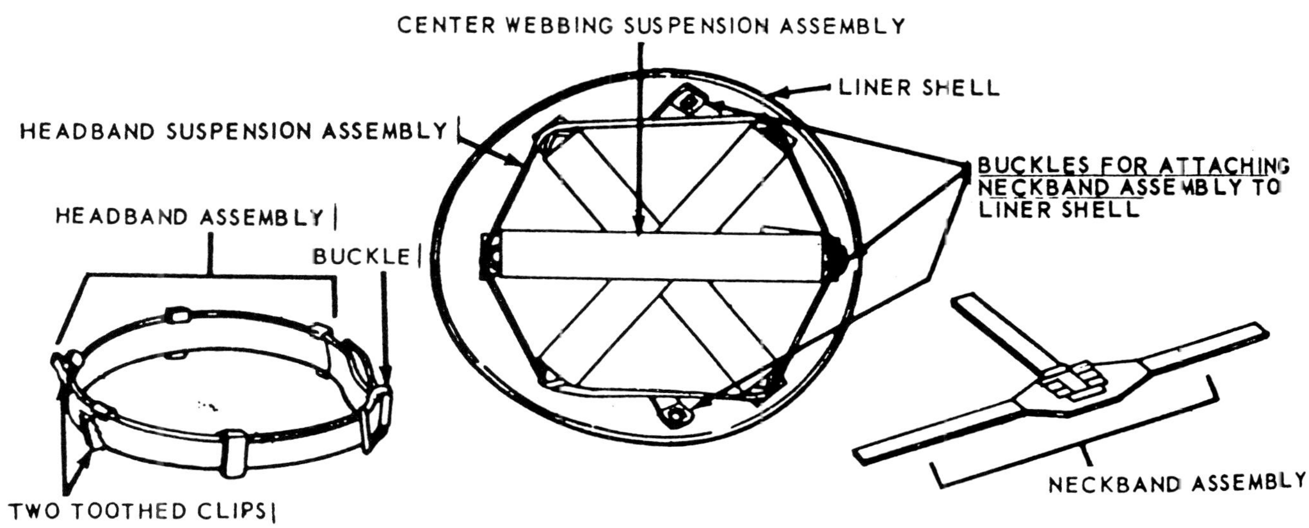

B. INTERIOR (COMPONENTS NOT ASSEMBLED)

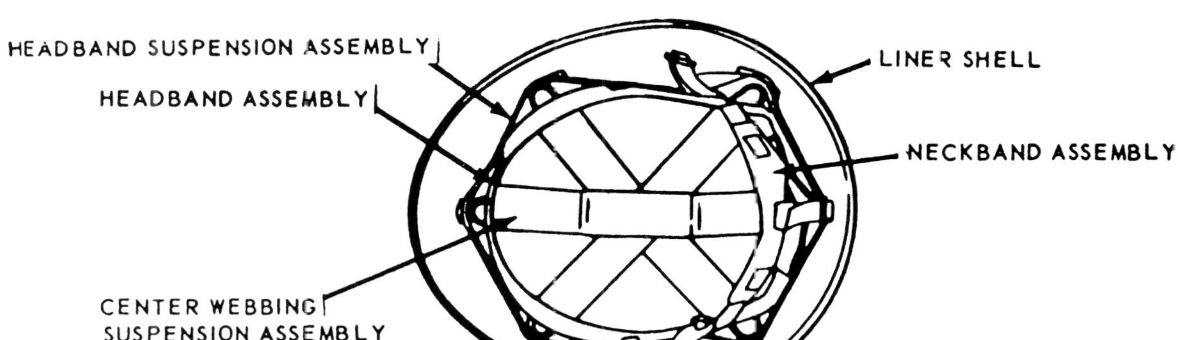

C. INTERIOR (COMPONENTS ASSEMBLED)

ME 8400-201-23/16-3 C3

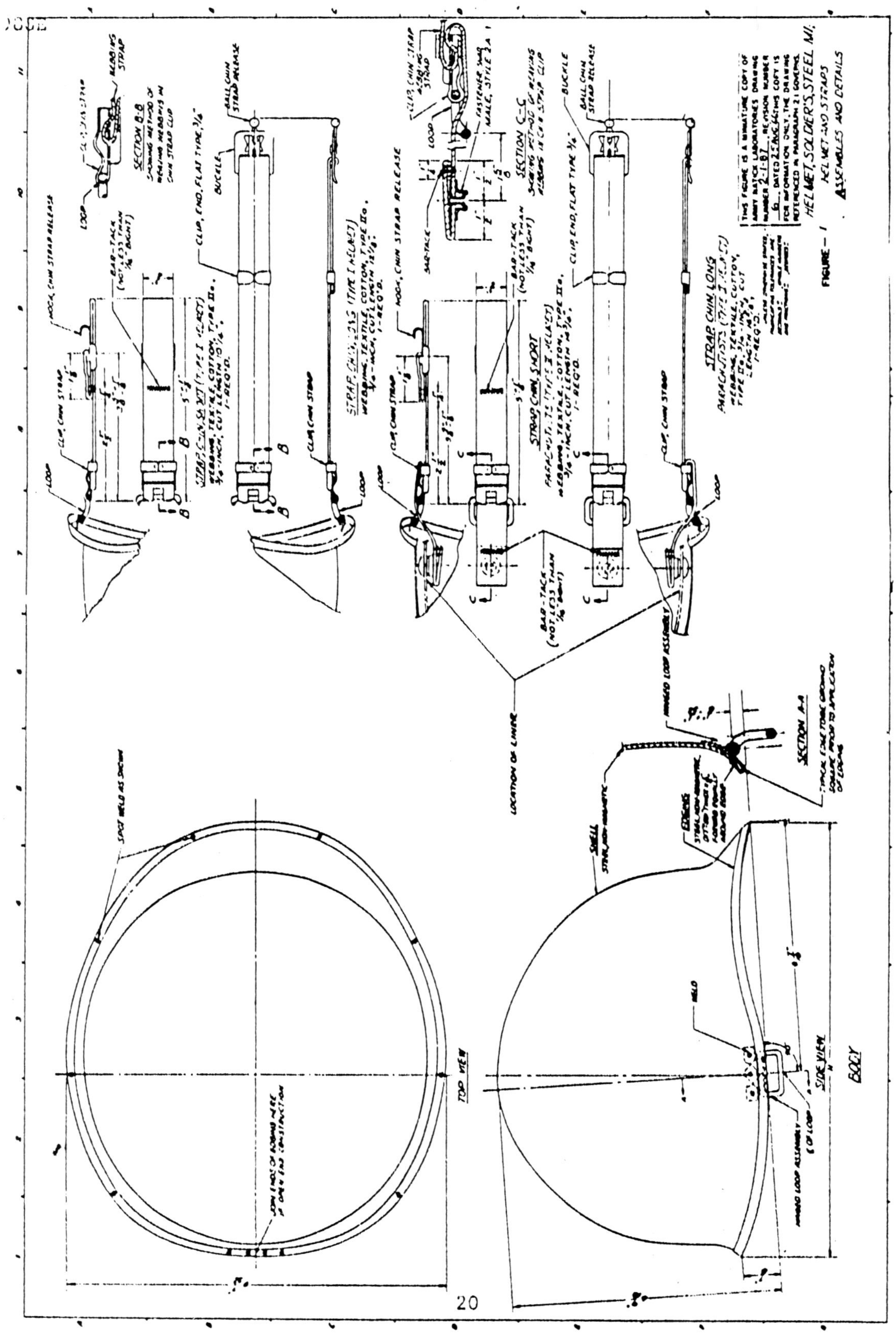
HELMET, SOLDIER'S, STEEL M1
HELMET AND STRAPS
ASSEMBLIES AND DETAILS
FIGURE - 1
TOP VIEW
SIDE VIEW
BODY
SECTION A-A
SECTION B-B
SECTION C-C
LOCATION OF LINER
20

MIL-H-41802B

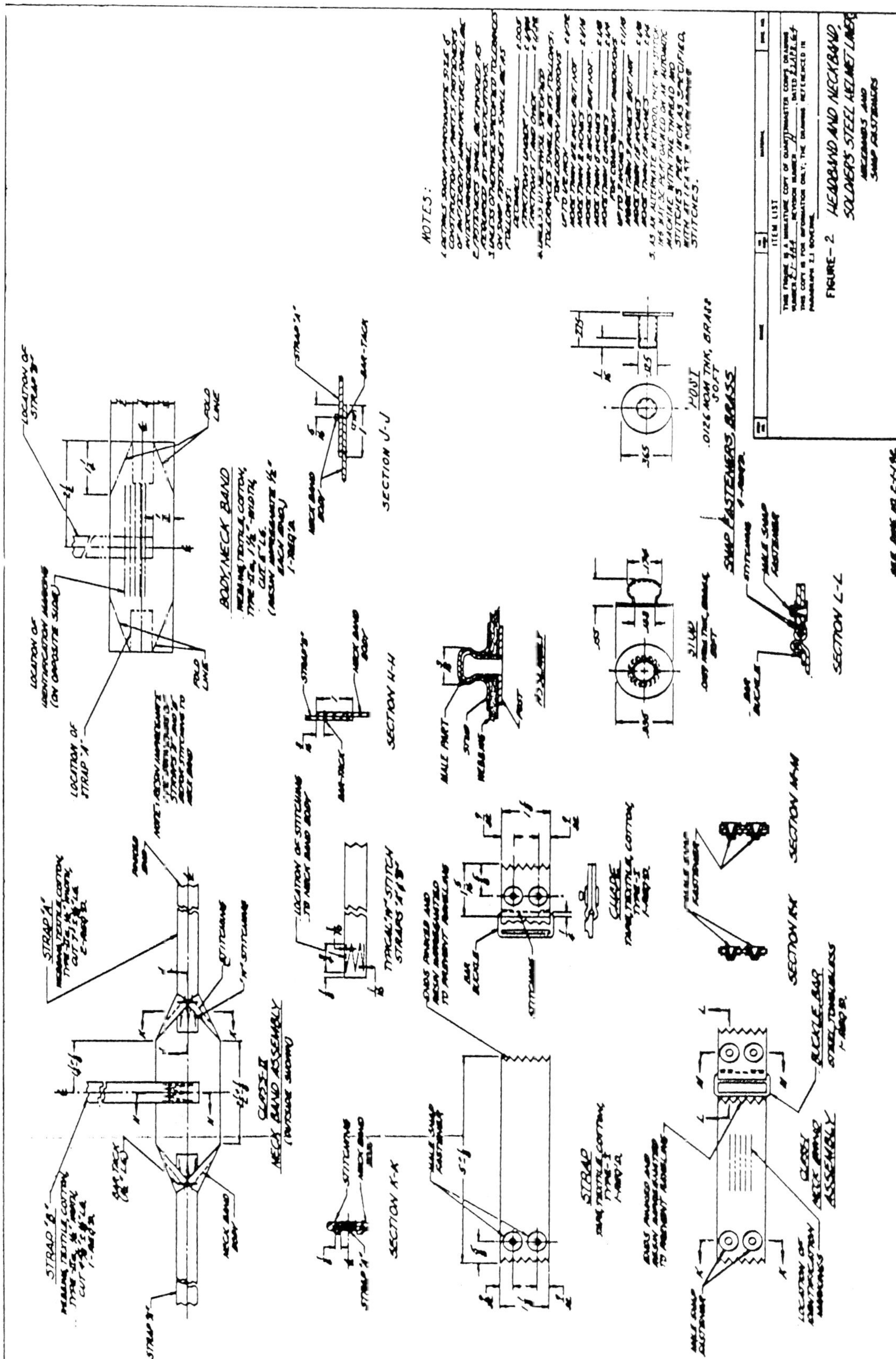
NOTES:
FIGURE-2
HEADBAND AND NECKBAND, SOLDIERS STEEL HELMET LINER
ITEM LIST
SECTION J-J
SECTION H-H
SECTION K-K
SECTION L-L
SECTION M-M
BODY, NECK BAND
STRAP "A"
STRAP "B"
STRAP
BUCKLE BAR
SNAP FASTENERS, BRASS

MIL-H-41802B

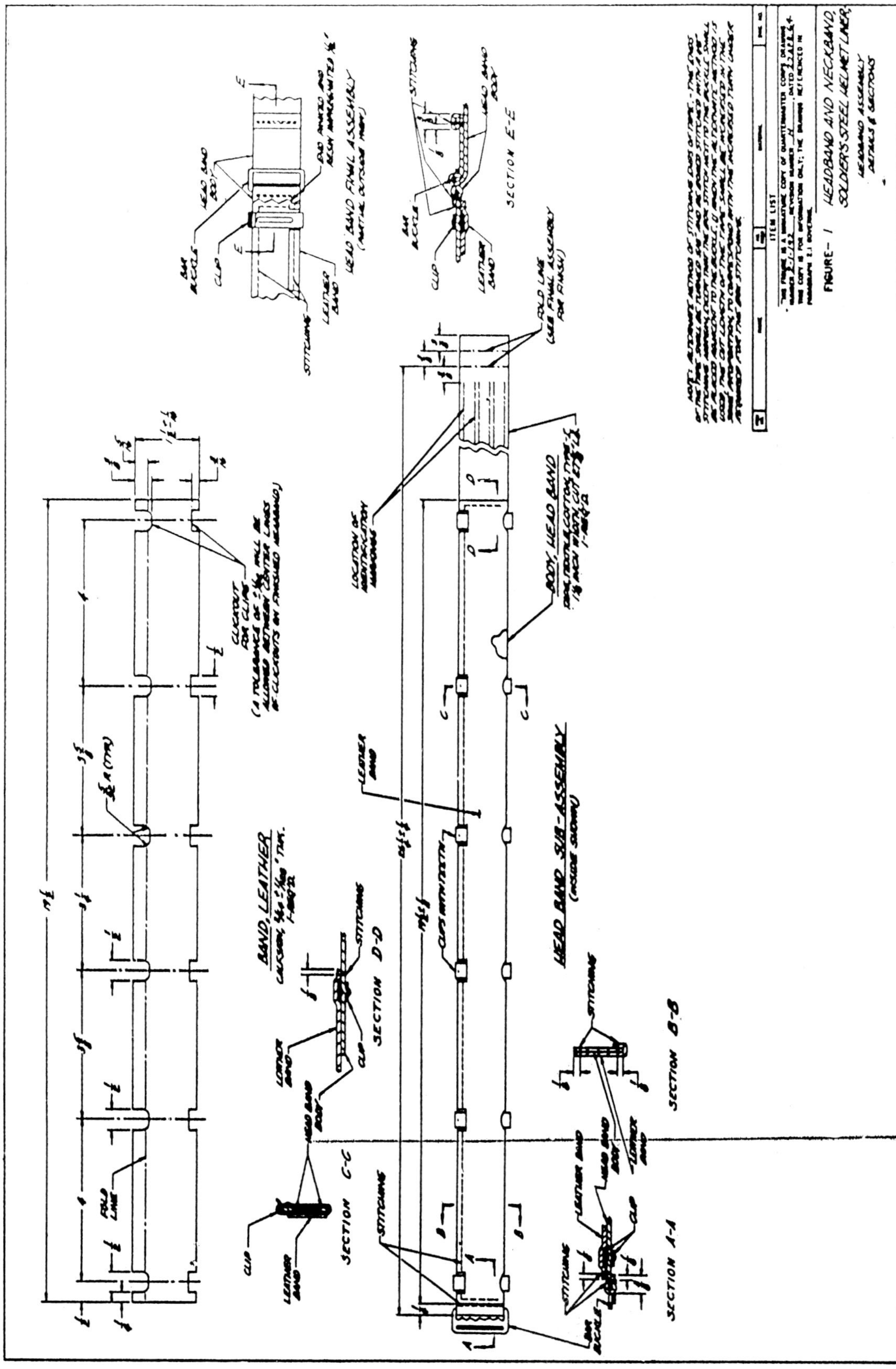
SECTION A-A
SECTION B-B
SECTION C-C
SECTION D-D
SECTION E-E
BAND, LEATHER
BODY, HEAD BAND
HEAD BAND SUB-ASSEMBLY
(INSIDE SHOWN)
HEAD BAND FINAL ASSEMBLY
FIGURE- I
HEADBAND AND NECKBAND,
SOLDIERS STEEL HELMET LINER,
HEADBAND ASSEMBLY
DETAILS & SECTIONS

MIL-H-43178A

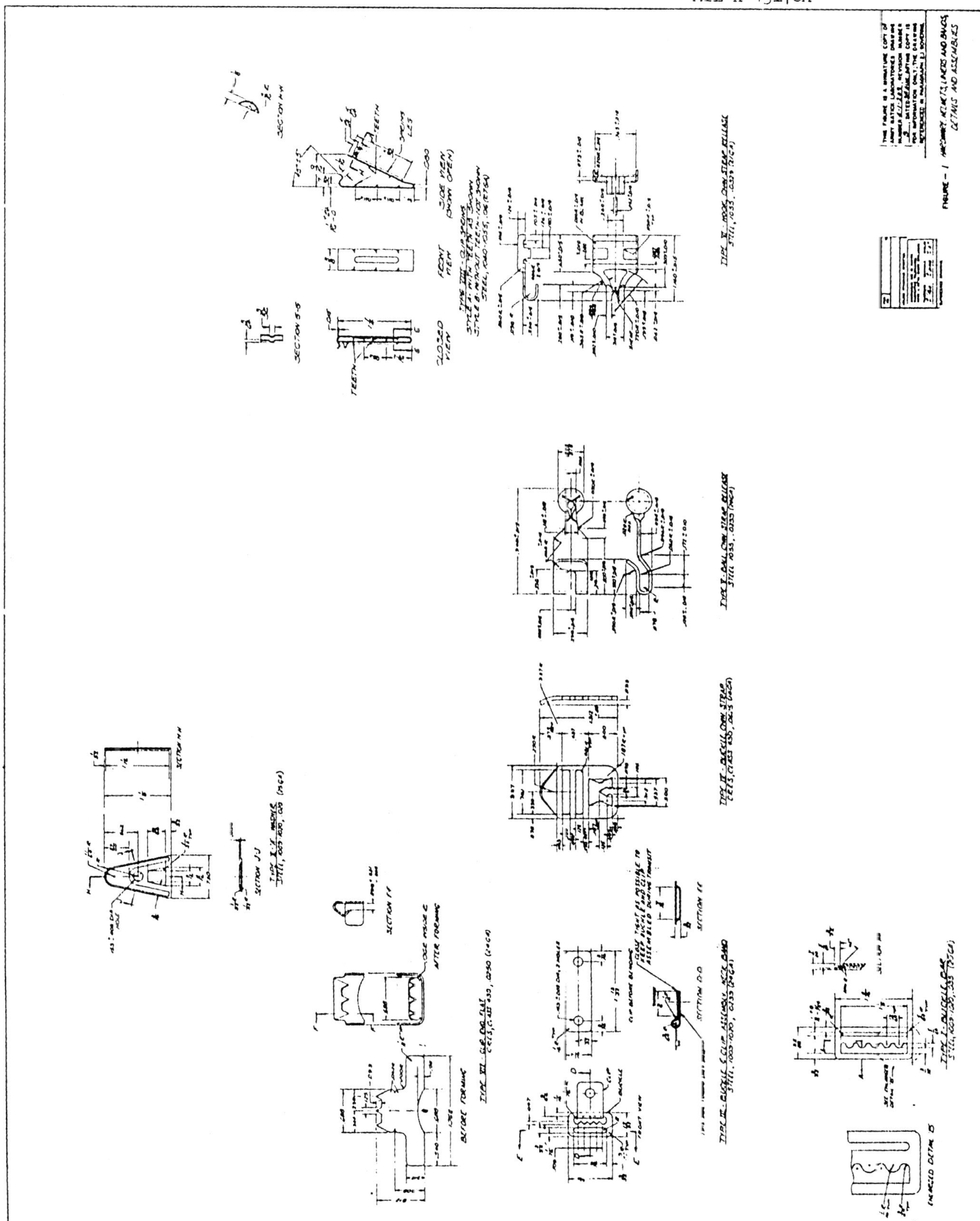

11.

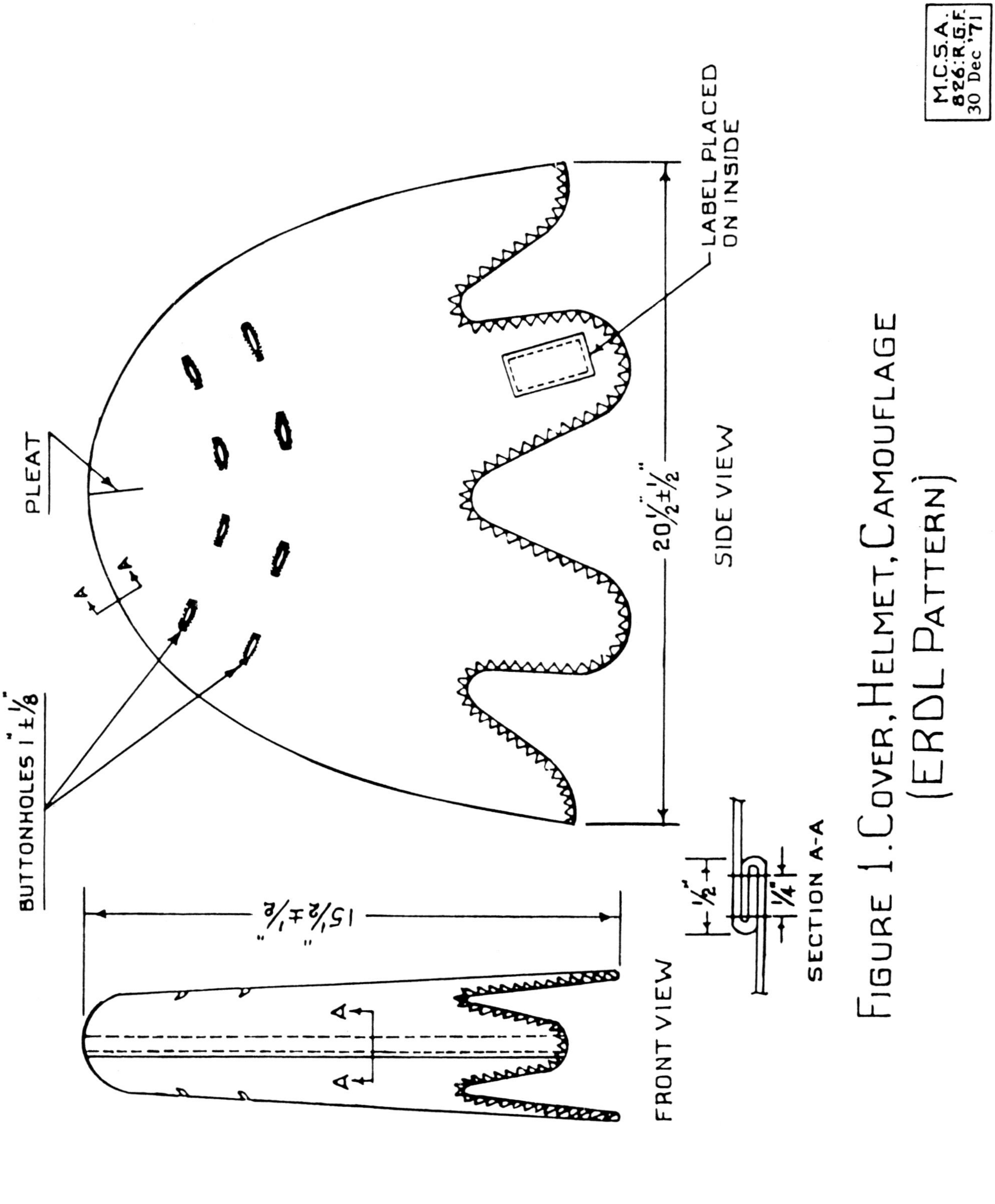

FIGURE 1. COVER, HELMET, CAMOUFLAGE (ERDL PATTERN)

MIL-H-1988F

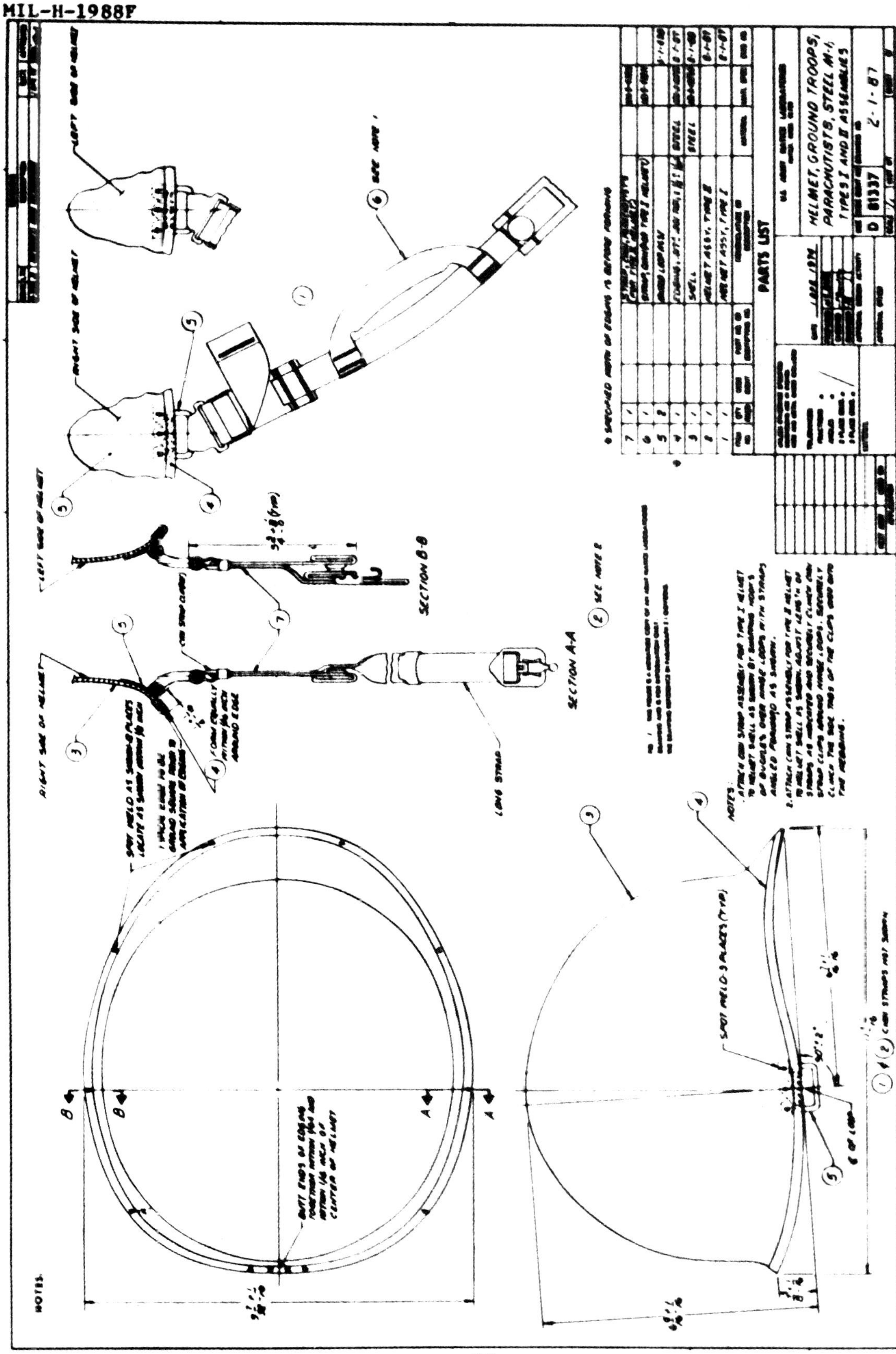

22

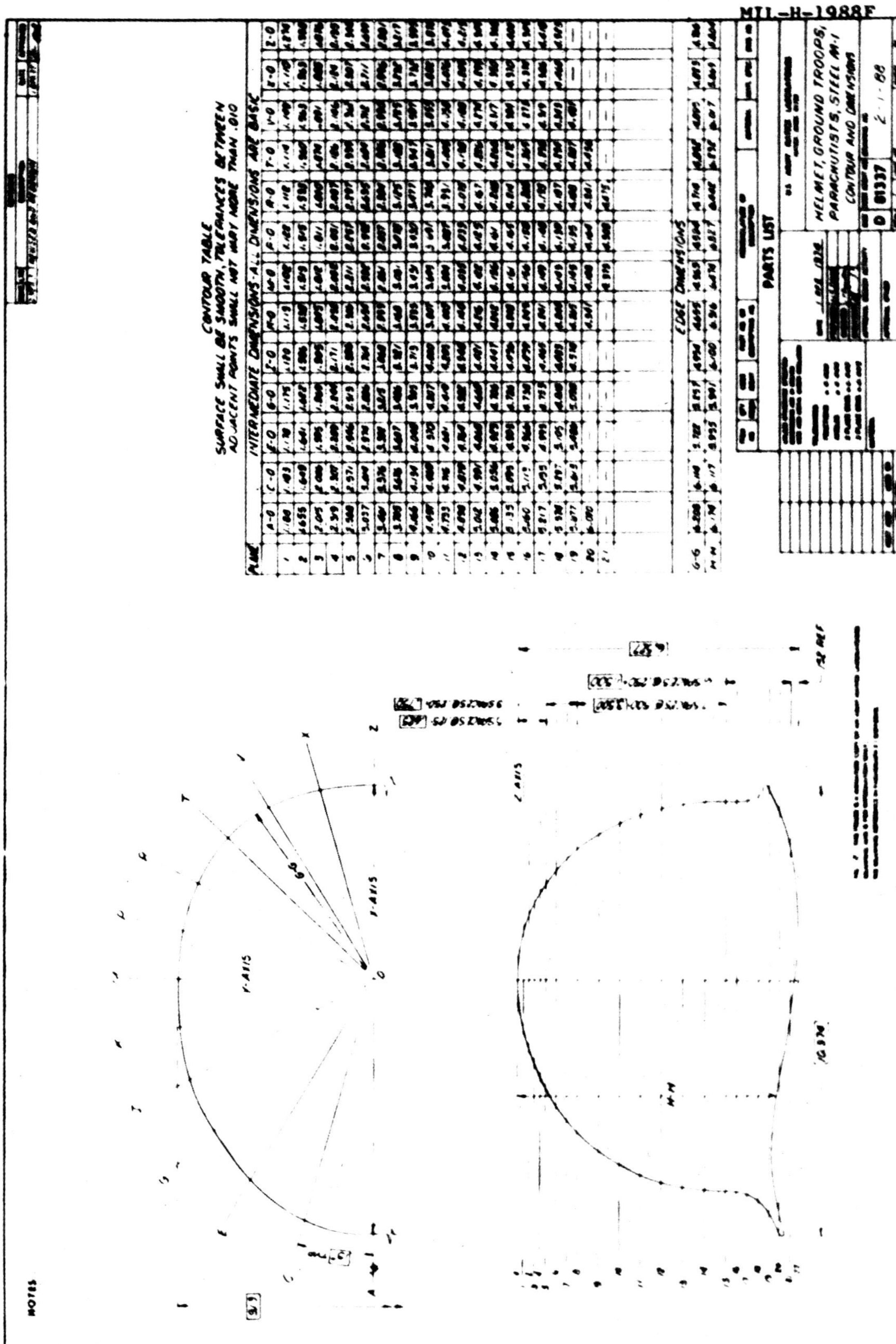
MIL-H-1988F
CONTOUR TABLE
SURFACE SHALL BE SMOOTH. TOLERANCES BETWEEN ADJACENT POINTS SHALL NOT VARY MORE THAN .010
EDGE DIMENSIONS
PARTS LIST
HELMET, GROUND TROOPS, PARACHUTISTS, STEEL M-1 CONTOUR AND DIMENSIONS
NOTES

23

MIL-H-1988F

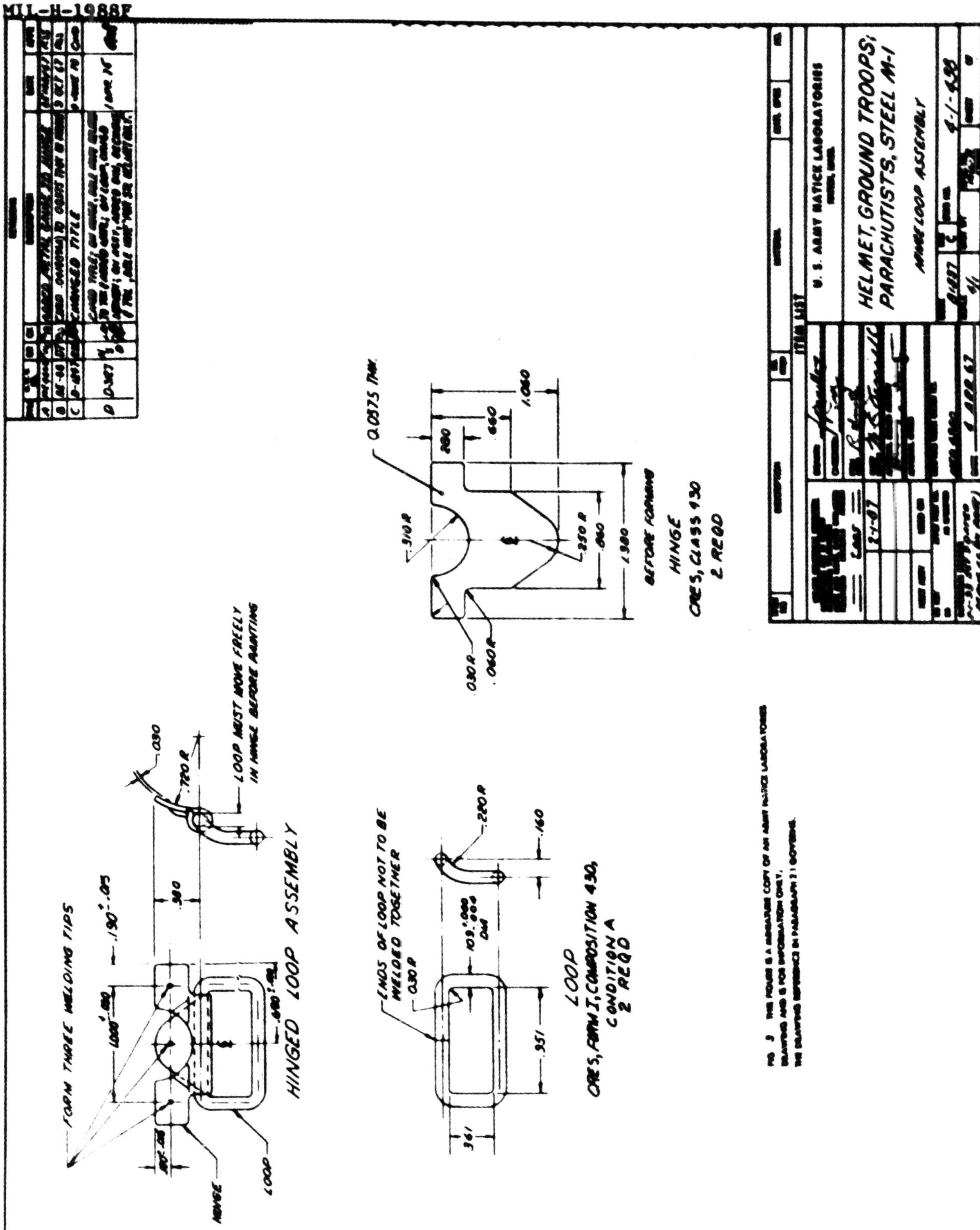

24

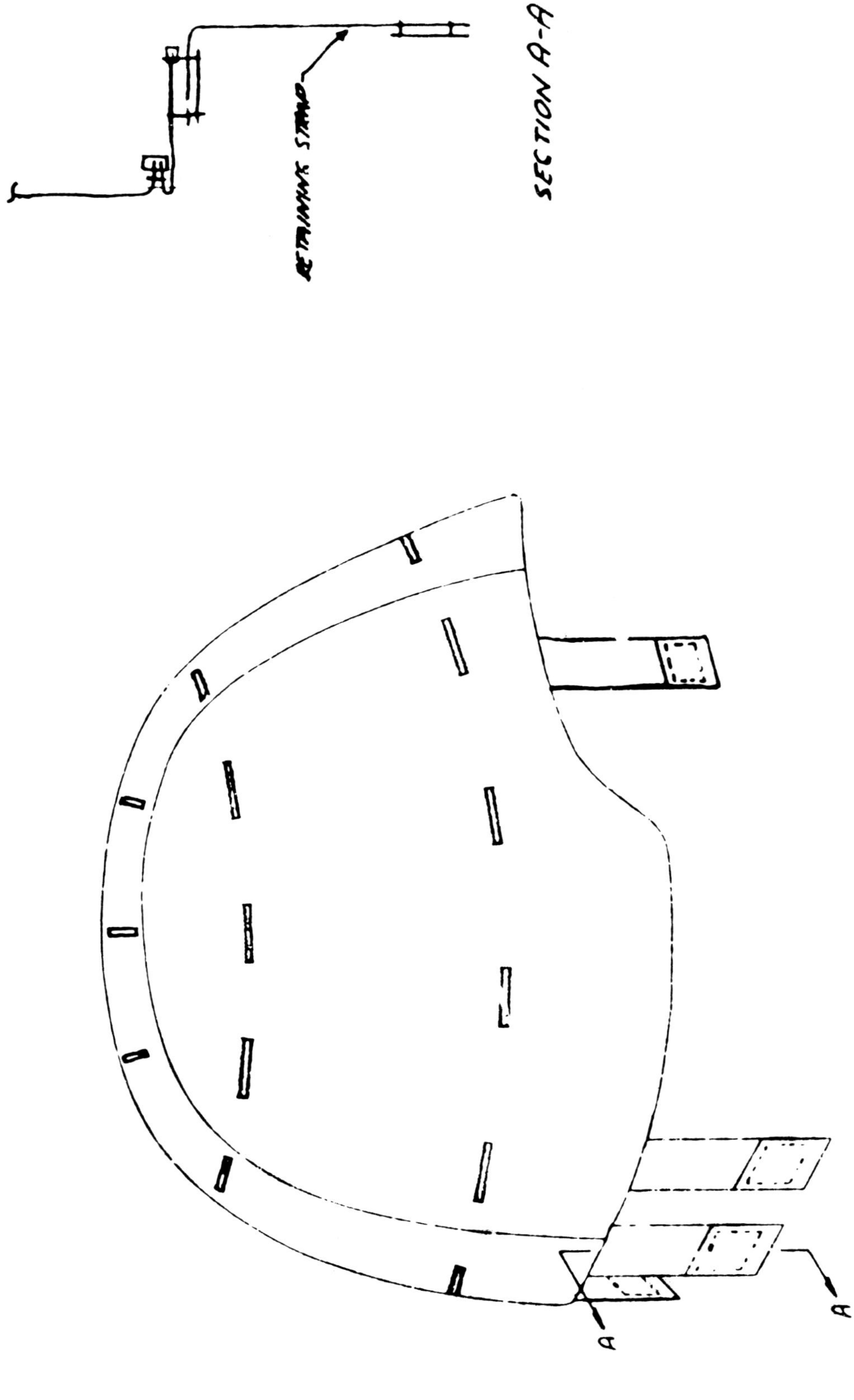

FIGURE 1 - COVER, GROUND TROOPS - PARACHUTISTS HELMET

Also from the publisher

THE M-1 HELMET: A HISTORY OF THE U.S. M-1 HELMET IN WORLD WAR II

Mark A. Reynosa

This new book presents over seven years of research into the history of the M-1 helmet during World War II, and provides the most comprehensive examination of its development and production. All aspects of M-1 helmet production are covered including: the helmet body, the fiber liner, the plastic liner, the parachutist helmet, helmet camouflage, helmet modifications, helmet paint schemes, and toy helmets. Every production helmet version is presented in full color photographs, including detail shots and production markings. Also included are World War II era photographs of the helmet samples, helmet production, and helmets worn in training or in action. This book is a valuable reference to both historians and collectors.

Size: 8 1/2" x 11" over 350 color and b/w photographs

112 pages, hard cover

ISBN: 0-7643-0074-1 $39.95